Emotion in Farmed Animals

The emergence of scientific data confirming the capacity of nonhuman animals to feel what is happening to them and experience positive and negative emotions has created an uncomfortable moral dilemma for many humans. To meet the demand for animal products in the 21st century, millions of animals are confined and suffer in industrial farm settings.

Acknowledging that a sense of helplessness to end this entrenched suffering on such a large scale can lead to inertia, this book instead seeks to bring change through stories of individual animals now living on farmed animal sanctuaries. These rescued animals have narratives that reveal them as individuals with emotional capacity and a future. Each chapter looks at a particular species' entwined history with humanity, as well as the biological and neurological structures that unequivocally confirm their ability to experience positive and negative emotions. This aspect is brought to life through the stories of individual animals living the good life after rescue and rehabilitation.

Farmed animals all have stories to tell but few get the chance to do so; this book amplifies the voices of those who have been able to reclaim basic rights in a human-dominated world with their stories providing insight into the uniqueness of every pig, cow, bird, rabbit and more intended for our plates and others for companionship or sheer joy.

This book will be of interest to those studying or working in animal ethics, welfare and law; veterinary science; running a farmed animal sanctuary; or simply seeking to understand more about the animals with whom we share this planet.

Emotion in Farmed Animals

Ethical Challenges for Animal Welfare

Vicki Hutton

CRC Press
Taylor & Francis Group
Boca Raton London New York

CRC Press is an imprint of the
Taylor & Francis Group, an **informa** business

Designed Cover Image: Noelle the pig, Preena the turkey, Harriet the goat and Chevvy the sheep from Edgar's Mission, Australia. Photographs provided by Rebecca Hutton and reproduced with permission.

First edition published 2026
by CRC Press
2385 NW Executive Center Drive, Suite 320, Boca Raton FL 33431

and by CRC Press
4 Park Square, Milton Park, Abingdon, Oxon, OX14 4RN

CRC Press is an imprint of Taylor & Francis Group, LLC

ISBN: 9781032855844 (hbk)
ISBN: 9781032848600 (pbk)
ISBN: 9781003518853 (ebk)

DOI: 10.1201/9781003518853

Typeset in Minion
by codeMantra

Dedication

For my children who share my love and respect for all animals, for my parents who instilled in me a deep and lasting desire to protect animal welfare, and for all the animals who have allowed me a glimpse into their world.

Contents

About the author

Vicki Hutton is passionate about animals and their welfare, having lived with, worked with, observed and written about animals for many years. After starting her career in the health and community sectors as a mental health therapist, crisis counsellor and group facilitator, Vicki moved into education and research while completing her doctorate in Applied Psychology. She has published widely in the human–animal relationship field, as well as in the areas of mental health, stigma and discrimination, and social cohesion. She currently lives with an assortment of animals who all contribute in their unique ways to her writing.

Acknowledgements

This book highlights the uniqueness of farmed animals when they are able to safely express their personalities and emotional capacity. Bringing together all these stories reveals both the possibilities of a new future for farmed animals and some of the harms that humans have inflicted on these sentient animals over the years. Without the contributions of the animals and humans from 23 different farmed animal sanctuaries around the world, this book would not have been possible. Sincere thanks to:

Agia Marina Donkey Rescue in Greece
Austin Farm Sanctuary in the USA
Black Sheep Animal Sanctuary Ōtaki in New Zealand
Clement Rabbit Rescue in Wales
Edgar's Mission in Australia
Flow Farm and Sanctuary in China
Forever Friends Animal Rescue in Australia
Furever Farm in Australia
Galahad's Animal Sanctuary in Australia
Goodheart Farm Animal Sanctuary in the United Kingdom
Greyton Farm Animal Sanctuary in South Africa
Horse and Farm Animal Sanctuary in Wales
Ippoasi Animal Sanctuary in Italy
J&D's Farm Sanctuary in Australia
Lefty's Place Farm Sanctuary in Australia
Magzalea Farm Sanctuary in the USA
Mother Clucker's Microsanctuary in the USA
Mother's Heart Sanctuary in Indonesia
Saint Beans Animal Sanctuary for Lambs in New Zealand
Strong Hearts Farm Sanctuary in Australia
The Flicka Foundation Donkey Sanctuary in the United Kingdom
Vrouva Farm in Greece
Woodstock Farm Sanctuary in the USA

Introduction

For some people, the decision to stop eating meat and animal products happens slowly. Like a persistent tune stuck on repeat, the uncomfortable connection between living animals and meat on a plate grows increasingly persistent despite all efforts to block it.

For others, there is a moment in time that marks this life-changing decision. It might be images of live animal exports, or pigs stuck in breeding crates, or chickens with distorted feet struggling to live and breathe in cages. My moment in time happened in the picturesque Lake District of Cumbria (United Kingdom), where a flock of black-faced Swaledale sheep browsed in a meadow below our hotel dining room window.

As my uncle prompted me to eat up a traditional English breakfast of devilled kidneys, my eyes were drawn to the living sheep, and I could not help but wonder when their turn would come to be slaughtered, mixed into a spicy sauce and served up on a fine bone China plate. My uncle, a Yorkshire farmer and auctioneer, seemed genuinely puzzled when I made this connection, but solved the problem by directing me to the bacon and eggs that could be eaten out of sight of pigs or chickens.

My fork hovered over the chopped up spicy kidneys that looked nothing like the sheep and their cheeky lambs so happy to be alive on a sunny morning in the Lake District. The uncomfortable connection had been made and could no longer be ignored, all thanks to a flock of black faced sheep and a plate of devilled kidneys. Over the years, I learned that it is not always easy to reject animal flesh and animal products in a human-dominated world. Change is coming, but not fast enough for the millions of animals who continue to suffer and die in order to satisfy human demands.

My Yorkshire family, who had opened their home to me during my extended visit, did their best to understand this sudden refusal to eat the meat, eggs and dairy products that were staples in every hearty meal. They listened with tolerant mirth when I tried to explain the connection I perceived between these products and Sally the cow, Jasper the pig and Hettie the hen—friends I had made and named on my arrival at the farm. They tried to accommodate their Australian relative's passing phase, but failed to understand that picking meat out of a stew and serving a plate of the remaining vegetables in gravy would not work and that eggs and milk were not ours to take.

Growing up surrounded by companion animals had left me with no doubt that animals have feelings, but it was not until I met and named Sally, Jasper, Hettie and the many other "beasts" on my uncle's farm that I was able to expand my recognition beyond the limited animals to which most urban-dwelling humans are exposed. The simple act of naming these individuals shifted them from a food

DOI: 10.1201/9781003518853-1

source to sentient beings, just as watching the black-faced lambs frolicking, while portions of their unlucky relatives congealed on my breakfast plate had solidified this perception.

Several decades later in Cambridge, over 400 kilometres from the picturesque Lake District, the *Cambridge Declaration on Consciousness* was signed at the Francis Crick Memorial Conference on Consciousness in Human and Non-Human Animals. The 2012 Declaration affirmed that humans are not unique in possessing parts of the brain complex enough to support conscious experiences, with convincing evidence that many animals have neuroanatomical structures and neurochemical pathways fundamental to core emotions. They may not express these emotions in a human-centric way, but that in no way negated their existence.

Despite this formal acknowledgement that some animals have the capacity to experience feelings, many farmed animals continue to be deprived of their freedom and forced to live in industrial farm settings. Acknowledging emotional capacity in these animals means relinquishing learned attitudes of animals as commodities and dealing with the resultant cognitive discomfort when confronted with the ongoing plight of factory-farmed animals.

The animals on my uncle's Yorkshire farm had relative freedom and were never subjected to the intensive animal agriculture that came to dominate industrialised countries to meet the insatiable demand for meat and animal products. They never experienced the mass confinement and slaughter endemic in a human world. The chickens did not experience rapid growth beyond the capacity of their legs to support, nor were the bobby calves taken from their grieving mothers within hours of birth. But even the "good life" on my uncle's farm ended with premature death—the cows and pigs were sent to auction, and many of the chickens ended up on plates.

Naming the animals on my uncle's farm made it impossible for me to eat them, just as naming my companion animals had turned them into family members. Farmed animal sanctuaries, which are dedicated to the rescue, rehabilitation or rehoming of farmed animals, learned the value of naming an animal long ago. All around the world, a growing number of sanctuaries collect and shelter those animals whose life trajectory is dramatically altered after escaping their human-determined fate. A name bestows individuality and identity to the downed sheep that someone noticed and rescued from an icy paddock, or the spent hen who fled her human captors and certain death when no longer able to produce an unnatural supply of eggs. Rescuing and naming the tiny lamb found bruised and bleeding on the roadside after slipping through the cracks of a live transport vehicle where he was born hours earlier to a mother destined for death gave him a future. And rescuing and naming the male calf born in the dairy industry, separated from his mother and scheduled for an early death as a waste product of the dairy industry, validated his worth as a sentient being. These outcasts who are lucky enough to find themselves in farmed animal sanctuaries gain a name, an identity and a future.

A growing number of sheep, pigs, cows, goats, chickens, roosters, turkeys, ducks, geese, rabbits, donkeys, camels, llamas and alpacas, all destined to service

Figure I.1 Farmed animals have the capacity to experience feelings. (Photograph from Edgar's Mission provided by Rebecca Hutton and reproduced with permission.)

humans in some way, have found themselves swept up in a global grassroots movement to recognise and respond to them as sentient beings. That global movement offers them a future and a voice as they become ambassadors of change.

In 1974, philosopher Thomas Nagel asked the simple question, "What is it like to be a bat?" (p. 25). Ecologist and author Carl Safina (2015) made a similar observation when he stated, "My task now—a much harder task, a much deeper task—would be to endeavour to see *who* animals simply *are*—like us or not" (p.13). It can be hard to see who animals are when confronted by horrific images of animal death and suffering on a massive scale in the industrialised world of animal agriculture. Psychic numbing, a flawed protective mechanism, can create a sense of indifference and removal of personal responsibility when stories and images become too horrific and widespread to remedy. Images of sheep living and dying on live export vessels can induce helplessness to enact change on this large scale, but one tiny lamb found injured on the side of the road personalises the tragedy to a more manageable scale. Give that lamb a name and a story, and it is possible to break through human psychic numbing and start the process of change.

The animals in farmed animal sanctuaries all have stories to tell. Many have borne the brunt of humanity's cruelty and yet retained an innate goodness that allows them to forgive and regain joy in life. Having won the battle to reclaim basic rights in a human-dominated world, their stories provide insight into the uniqueness of residents in these sanctuaries. Some of their stories are captured in the following chapters in an attempt to show what it means to be a sheep, cow, goat, chicken and more if given the opportunity. The long and entwined history of their ancestors before and after human intervention is juxtaposed against the possibilities inherent in a new life free from the physical and emotional constraints of human captivity as each chapter delves into the lives of farmed animals, past, present and future.

Farmed animals have feelings and voices, something that is increasingly being recognised as humans confront the far-reaching implications of ingrained attitudes and behaviours in a shared world. This book celebrates the lives and voices of some unique individuals who personalise the battle to change the fate of all farmed animals currently being waged on a global level.

REFERENCES

Cambridge Declaration on Consciousness. (2012, July 7). Written by Philip Low and edited by Jaak Panksepp, Diana Reiss, David Edelman, Bruno Van Swinderen, Philip Low and Christof Koch. University of Cambridge.

Nagel, T. (1974). What is it like to be a bat? *Philosophical Review, 83*, 435–450.

Safina, C. (2015). *Beyond words: What animals think and feel*. Henry Holt and Company, LLC.

1

Welfare, ethics and productivity: An incompatible mix?

There was a time when a traditional farm animal was deemed to be living the good life if they were not sick or injured, they were provided with regular food and water and they had access to adequate shelter. Even when they transitioned into "factory farmed" animals living in industrial farm systems, there remained a perception that satisfying those basic requirements was sufficient to meet legal and welfare requirements. However, the publication of United Kingdom animal welfare activist Ruth Harrison's book *Animal Machines* in 1964 set in motion a chain of events that would have repercussions well into the twenty-first century.

Harrison's book opened with a clear statement of intent:

> I am going to discuss a new type of farming, of production line methods applied to the rearing of animals, of animals living out their lives in darkness and immobility without a sight of the sun, of a generation of men who see in the animal they rear only its conversion factor into human food.
>
> *(p. 35)*

This was followed by a series of probing questions that remain unanswered in many sectors of contemporary society:

> How far have we the right to take our domination of the animal world? Have we the right to rob them of all pleasure in life simply to make more money more quickly out of their carcasses? Have we the right to treat living creatures solely as food converting machines? At what point do we acknowledge cruelty?.
>
> *(p. 37)*

DOI: 10.1201/9781003518853-2

Harrison was referring to intensive animal agriculture, also known as factory farming or industrial farm systems, where animals are kept at high density and subject to stressors associated with unnatural confinement. Whether caged and unable to move freely, forcibly separated from their newborn or injured and sick, factory farmed animals can suffer in many ways during their shortened lifespan. With some estimates that around three-quarters of farmed animals continue to live in industrial farm systems, this places the welfare of many animals at serious risk (Anthis & Anthis, 2019). Global statistics indicate that animal slaughter—especially among small animals such as chickens and ducks—has consistently outpaced human population growth, highlighting the need to address the treatment of these sentient beings (orzechowski, 2024).

Domestication of animals has been described as both biological and sociological, entailing physiological changes as well as dependency and domination (Cochrane, 2014). Many domesticated animals were substantially reshaped by humans in the process of domestication and industrial farming, raising questions about the moral responsibility to ensure the wellbeing of animals changed so dramatically to fulfil human needs (Edison & Esvelt, 2020). Selecting for appearance or behaviour does not automatically translate into wellbeing, nor does enhanced reproductive performance necessarily promote improved health. It is no longer possible to disentangle animal agriculture from animal welfare science and ethics, given the direct and ongoing impact of human manipulations on farmed animals.

The 1970s saw a number of books on animal ethics and advocacy emerge, including *Animal Liberation* (Peter Singer, 1975) and *Slaughter of the Innocent* (Hans Ruesch, 1978). As Harrison's message reverberated through animal advocacy movements, this trend continued into the 1980s with the publication of pivotal books such as *The Case for Animal Rights* (Tom Regan, 1983), *Animal Rights and Human Morality* (Bernard Rollin, 1981) and *Animals and Why They Matter* (Mary Midgley, 1983).

American philosopher Tom Regan summed up the plight of farmed animals with the succinct quote:

> To treat farm animals as renewable resources is to fail to treat them with the respect they are due as possessors of inherent value. It is, then, to treat them unjustly, and this finding makes all the difference in the world when, as we are supposing, the farmer claims that he is 'within his rights' in raising these animals for human consumption [...] Morally this practice ought to cease, and consumers ought to cease to support it.
>
> *(1983, pp. 345–346)*

Philosophical and ethical discourses were soon reinforced by research papers, scientific journals and more books as the field of animal welfare science gained momentum in the 1990s and 2000s. Many lessons would be learned in those early, heady days of animal liberation. In particular, as long as people believed they needed meat and animal products to survive, intensive animal agriculture was the only way to produce them economically and in sufficient quantities to satisfy the relentless demand. Human desires combined with financial imperatives provided an iron-clad rationale to continue confining and slaughtering billions of animals and birds every year, despite evidence of unimaginable atrocities behind the closed doors of industrial farm systems.

As sentient creatures, animals should have the right to "someone" status rather than "something" that is nothing more than property, a notion that is seemingly supported by findings from the Sentience Institute's *Animals, Food, and Technology Survey: 2021 Update* (Ladak & Anthis, 2022). If almost three-quarters of responders reported discomfort towards the animal farming industry, and around half expressed relatively strong support for banning both slaughterhouses and factory farming, the obvious question is how these systems continue to exist. Human psychology can help explain the mismatch when a person claims to like, even love, animals, while they simultaneously accept certain animals as commodities for use and abuse.

In 1957, social psychologist Leon Festinger published a book entitled *A Theory of Cognitive Dissonance* in which he described the way people deal with inconsistencies between what they think or believe and the way they behave. The greater the difference, the greater cognitive dissonance they experienced, manifesting as psychological tension and discomfort and the need to relieve it. Eating animal products exemplifies this tension when people say they are animal lovers, but find reasons to continue eating animal flesh and products.

Another psychologist, Melanie Joy (2009), described the power of normalisation, which goes some way to easing the tension. Joy suggested the "three Ns theory of carnism"—that meat eating was necessary, natural and normal—and this provided sufficient validation to continue eating meat. Researchers Jared Piazza and colleagues (2015) added a fourth N to the mix—eating meat is nice (pleasurable and tasty). That made it simpler to continue eating meat, while overlooking the suffering and avoiding cognitive dissonance. Industrial farm systems are distanced from urban dwellers, and slaughterhouses are often protected by ag-gag laws, thus allowing the plastic wrapped meat, dairy products and eggs in supermarkets to be dissociated from the abuse and slaughter of the animals from whose bodies they originated.

Underpinning these barriers is speciesism, a disputed but deep-rooted concept that largely comes back to the same circular argument: speciesism assumes that humans are superior to nonhuman animals by virtue of being human (Jones, 2020). This self-appointed moral superiority validates the right to exploit animals, despite being built on the shaky foundation of unsubstantiated human supremacy.

Recognising these obstacles may be why Tom Regan commented during an interview about veganism and animal rights in 2013 that:

> The world is not going to change overnight, nor in my lifetime, nor in my grandchildren's lifetime. But it will change, if we change, not for a day or a year, but for a lifetime. Yes, it will change, if we change.
>
> *(all-creatures.org, 2013)*

THE FIVE FREEDOMS OF ANIMAL WELFARE

Meanwhile, Ruth Harrison's powerful descriptions of abhorrent treatments imposed on countless livestock, and poultry had caught the attention of the British government, which set up a committee to investigate the welfare of farmed animals in general. By 1965, Professor Roger Brambell and the Committee released an 89-page report with the clear statement that "…an animal should at least have sufficient freedom of movement to be able without difficulty, to turn round, groom itself, get up, lie down and stretch its limbs" (p. 13). This was in addition to the basic freedoms to not feel hungry, thirsty, ill or in pain, thus beginning the evolution of animal welfare from a simplistic prevention of suffering to the active promotion of an animal's wellbeing, a nuanced distinction that could make a world of difference to the individual animal.

The report succinctly summarised the expectations for every animal in modern, intensive animal production methods: they should have ready access to fresh water and a diet to maintain health; an appropriate environment including shelter and a comfortable resting area; prevention, or rapid diagnosis and treatment of disease and injury; space and facilities to be with their own kind; and the aggregate of these freedoms ensuring the avoidance of mental suffering, fear and distress. Professor Brambell's Five Freedoms conception of animal welfare provided a catalyst to interrogate what really constituted animal welfare in a human-dominated world.

While the above freedoms provided descriptive markers to consider, the report provided only an implicit definition of farmed animal welfare against which to consolidate this information (Mellor, 2017). Over the years, other researchers became more explicit in their definitions by linking the state of the animal to their attempt to cope with their environment, physically and mentally. For example, a definition from ANSES, the French Agency for Food, Environmental and Occupational Safety, clearly stated that an animal's welfare should include their positive mental and physical state in relation to the fulfilment of their physiological and behavioural needs, as well as expectations. The animal's perception of a situation would thus impact their state (Ferreira et al., 2021). As these definitions edged closer to requiring the individual animal's perspective be considered to properly assess welfare, a dilemma arose around how to access and interpret the various animals' perspectives.

Abnormal behaviours offered one means by which to understand the negative affective states of animals and thus provide a window into their perspective. When human psychiatrist David Levy (1944) described the rhythmic rocking movements and other stereotyped behaviours shown by some emotionally disturbed children, he also drew attention to similar behaviours displayed by farmed animals. These included the repetitive weaving of horses confined in stables and the stereotyped head movements of caged chickens. Levy suggested that these abnormalities in both children and animals could relate to severely deprived environments, a notion that was of considerable relevance to Ruth Harrison's observations two decades later and the subsequent Brambell Report.

An early Welfare Quality tool, which arose from Brambell's Five Freedoms Model, synthesised scientific expertise with ethical considerations from stakeholders such as the general public, industry and political bodies. This tool focused on animal-based measures that were assigned scores in four key areas: good feeding, good housing, good health and appropriate behaviour (Littlewood et al., 2023). Overall, welfare was then determined by a combination of all four scores with the assumption that in the absence of any negative scores, the animal's welfare must be acceptable. However, the absence of abnormal behaviours alone could not be assumed to confirm animal wellbeing as these measurable observations only provided a starting point. A glaring omission was the animal's perspective, which would require a more nuanced understanding of an animal's reaction to specific features of their environment (Grandin, 2022). During the domestication process, many animals experienced changes to behaviour and physiology to emphasise specific traits such as placidity and rapid growth. Therefore, failure to detect disturbances in behaviour based on previously normal behaviours was no longer an accurate reflection of an animal's conscious subjective experience or suffering as sentient beings (Edison & Esvelt, 2020).

According to biologist Jonathan Birch (2021), whose testimony contributed to the United Kingdom's amendment to the *Animal Welfare (Sentience) Bill* (2021), sentient animals have feelings with a positive or negative quality. These included feelings of pain, pleasure, comfort, discomfort, boredom, excitement, contentment, frustration, anxiety and joy. John Webster (2007), a veterinarian and Emeritus Professor of Animal Husbandry, was more succinct when commenting that sentience was feelings that matter. This meant that no two sentient animals would have the same subjective experiences, confounding attempts to develop an accurate measure of animal welfare without taking an animal-centric, multidisciplinary approach that integrated neurophysiology, applied ethology, animal science, animal ethics, philosophy, animal law and animal communication (Mellor, 2017). Animal welfare could no longer be a matter of ticking off a checklist to meet legal, ethical and social requirements.

ANIMAL WELFARE AND FEELINGS THAT MATTER

As an academic discipline, animal welfare transitioned through several phases strongly linked to growing knowledge of animal sentience and conceptions of what it meant to achieve wellbeing. In human literature, it has long been recognised that health-related quality of life, as measured by the absence of symptoms, could not always equate with the more holistic concept of subjective wellbeing. Similarly, complying with the five freedoms and removing negative stimuli such as hunger, thirst and lack of space did not guarantee good welfare in terms of a meaningful life. When Tom Regan (1983) argued for the recognition of an animal's right to have a mental life that included perceptions, desires, beliefs, memories, intentions and a sense of future, it was clear that intensive animal agriculture could never meet these standards. Only the lucky few animals who escaped to farmed animal sanctuaries would ever be granted a sense of future.

Even the focus on humane slaughter championed by Dr Temple Grandin, designer of livestock handling facilities and a Professor of Animal Science, failed to address one unavoidable welfare concern. Ensuring that animals did not suffer, or suffered minimally at the point of death, disregarded the fact that farmed animals have no meaningful future beyond living and dying for human gratification.

Demands for greater animal protection gained considerable traction over the decades, but lacked a consistent global approach to welfare. Legislation often included "loopholes" open to interpretation or certain exemptions, such as allowances for farming practices that privileged the structures of modern animal husbandry within mass production farms, confined housing, high breeding and maturation processes, over animal welfare (Wacks, 2024).

When Ruth Harrison described the plight of farmed animals in 1964, these animals fell under the jurisdiction of the United Kingdom *Protection of Animals Act 1911*, which specifically stated:

> "Nothing in this section [...] shall apply—
>
> (a) to the commission or omission of any act in the course of the destruction, or the preparation for destruction, of any animal as food for mankind, unless such destruction or such preparation was accompanied by the infliction of unnecessary suffering.
>
> *(p. 2)*

Unfortunately, the objective definition of "unnecessary suffering" remained unclear, and even the 2006 updates to the 1911 Act seeking to clarify this concept further exemplified the ongoing tension between animal rights, animal welfare and human desires when basing unnecessary suffering on the outcome of a leading case Ford v. Wiley. Hawkins J:

> Where, however, the suffering was inflicted in the pursuit of a legitimate object, there must be proportionality between the suffering caused and the benefit sought. The necessity of the defendant's conduct is judged objectively, by the standards of a reasonably humane and caring person.
>
> *(Collinson, 2018, Paragraph, "What Constitutes Unnecessary Suffering?")*

Inevitably, questions must be asked as to whether eating, wearing or otherwise using animal products constitutes a "legitimate object", and how animal-loving individuals could justify the cruelties inflicted on farmed animals in the pursuit of personal gratification. This came under scrutiny in a 1997 legal case between McDonald's Corporation and McDonald's Restaurants Limited and Steel and Morris [1997] EWHC QB 366 ("McLibel"). McDonald's (United States) sued two individuals for defamation in relation to several claims published in leaflets that were distributed to the public. One claim was that McDonald's supported cruel farming practices, a claim that was partially rejected by the Court on the grounds that some of the practices were not cruel if performed correctly. These included the maceration of unwanted chicks, the manual debeaking of chicks, teeth clipping, tail docking, castration of pigs and the permanent indoor housing of battery hens, broiler chickens and sows (Voiceless, 2015, p. 28).

Globally, domestic laws failed to keep up with the relentless industrialisation of farming practices as the push to meet demand turned factory farms into a global phenomenon with limited and inconsistent protective legislation. Animals occupied a special place in most legal systems as "animate property" (Robertson & Goldsworthy, 2022, p. 256), which left them open to continued abuse and exploitation. As property, they were classified as something, not someone, with no personhood or voice in the legal system. Animal welfare legislation also had to include provisions for "human religious rites, cultural heritage and regional heritage" (p. 249), further confirming that anthropocentrism would always usurp the needs of animals.

The tension between animal protection, animal welfare and animal rights was challenged by the third phase of animal welfare conceptualisation, the emergent "welfare-aligned sentience" based on David Mellor's extension of the Five Freedoms model into the Five Domains model (Robertson & Goldsworthy, 2022). Welfare-aligned sentience relied on the critical need to recognise an animal's subjective responses to the positive and negative experiences, sensations, feelings and emotions that mattered to them.

THE FIVE DOMAINS MODEL OF ANIMAL WELFARE

The Five Domains model drew attention to a broader conception of welfare compared to Brambell's founding model of five freedoms. It touched on subjective experiences, specified areas of welfare concern and, importantly, provided practical advice on how the five designated areas for welfare improvement could be achieved (Mellor, 2017). While retaining an emphasis on removing negative indicators such as lameness, malnourishment and sensory impairment, it acknowledged an animal's cumulative subjective experiences as being pivotal to their wellbeing.

This observation becomes important when considering some of the animals that feature in the following chapters. Bruiser at Forever Friends Animal Rescue in Australia and Snoopy at Austin Farm Sanctuary in the USA are blind cattle, and Kiwi, the duck from Woodstock Farm Sanctuary in the USA, had mobility issues. Penny, the three-legged sheep at Strong Hearts Farm Sanctuary in Australia, received unconditional love and respect during her lifetime. Observing the day-to-day interactions of these so-called "disabled" individuals leaves no doubt that their joy in life remained strong surrounded by friends. While accusations of anthropomorphising their subjective experiences may arise, this can be countered by ecologist Carl Safina's logical question: if we recognise an animal is hungry or thirsty, why cannot we also recognise that they are happy or joyful when acting in that way? (Safina, 2015). Positive or negative physical, mental and emotional states are unique to each animal, thus positioning welfare as a quality within the animal rather than an abstract construct.

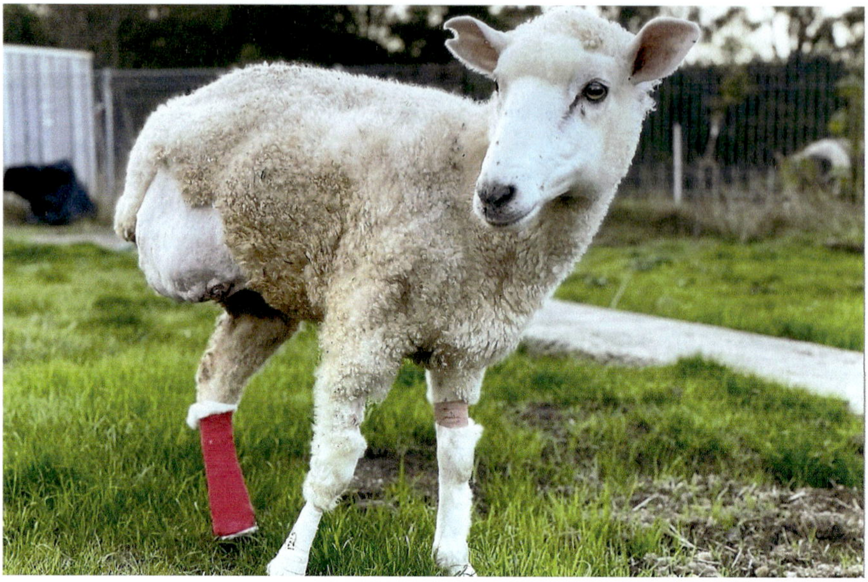

Figure 1.1 Penny, a much-loved sheep from Strong Hearts Farm Sanctuary in Australia. (Image supplied by Strong Hearts Farm Sanctuary and reproduced with their permission.)

Revised conceptual frameworks may seek to correct perceived errors or inadequacies in current ideas, but they also owe a debt to the earlier ways of thinking from which they evolved. Animal welfare scientist Professor David Mellor's 1994 formulation of the Five Domains Model had its origin in the highly influential Five Freedoms paradigm, but it was not merely an updated version. Its scope was far more encompassing. For almost three decades, the Five Freedoms model had become so integrated into the dimensions of animal welfare that it had become an implicit definition that outlined five targeted "freedoms" for welfare improvement and recommendations on how these freedoms could be achieved. Building on this, the Five Domains model recognised the need for both welfare compromise and welfare enhancement by drawing attention to the false positive that could arise when simply removing the negatives of the Five Freedoms model (Beausoleil et al., 2023). The Five Domains model thus applied a more pragmatic framework that recognised the negative impacts on animals could not always be removed and instead a relative grading was more realistic.

The Five Domains were conceptualised into five inter-related aspects of the animal's welfare state, with exemplars of both positive and negative impacts on the animal in each domain. Four physical/functional domains related to nutrition and hydration, physical environment, health and functional state and behavioural interactions. While adequate food and water in an industrialised farm setting is rarely an issue (although ducks may dispute this), extreme weather events and diseases such as Avian Flu and Swine Flu have increasingly raised questions about the safety of these environments for the health and functional state of confined animals. The exemplars included a range of qualitative and/or quantitative physical, physiological, pathophysiological, biochemical, immunological, neurological, and behavioural indicators, separated for wildlife and livestock.

Domain 4, renamed from Behaviours to Behavioural Interactions in the 2020 updated Five Domains model, recognised the impact of hindered or promoted agency when animals interacted with their environment, other nonhuman animals and humans. Affective states from these interactions were largely produced by the animal's brain processing sensory inputs from outside the body. In this way, Domain 4 differed from the preceding three domains that were determined by genetically programmed physiological mechanisms within the animal's body designed to maintain or restore homeostasis. Domain 4 introduced subjective actions of autonomy in response to situation-related factors and was particularly relevant to farmed animals who could be forced into unwelcome behavioural interactions (Mellor et al., 2020). Humans control all aspects of a farmed animal's conception, birth, life and death—including whether newborns can remain and bond with their mother, their living arrangements, their land or sea transport with non-group members and their method of death. The negative affective impacts of behavioural interactions were assessed on a continuum from "None" to "Very Severe", with positive affective impacts also assessed on this scale. Based on this information and in conjunction with the preceding three domains, an animal's potential mental/affective experiences relevant to their welfare state could be proposed for the fifth domain, Mental Experiences (Beausoleil et al., 2023). Domain 5 provided an assessment of the animal's overall affective experience and was deemed to equate to their welfare status.

Where an animal lacked agency in the human interactions of handling, transport and slaughter, the animal's fear-induced subjective perceptions and interpretations of their world not only shaped Domain 5, but could also impact physical functioning and productivity. This potential economic threat provided a catalyst to reconsider many of the handling and slaughter protocols, with Temple Grandin contributing significantly to information about livestock behaviour, design of facilities and humane slaughter. Unfortunately, humane slaughter and stunning practices were sometimes viewed through the anthropocentric lens of meat quality and human taste as much as animal welfare when considering the below descriptor appearing on Grandin's website:

> Stunning an animal correctly will provide better meat quality. Improper electric stunning will cause bloodspots in the meat and bone fractures [...] When stunning is done correctly, the animal feels no pain and it becomes instantly unconscious.
>
> *(Grandin, n.d.)*

Objectively interpreting an animal's subjective experience to complete Domain 5 could sometimes give rise to challenges and accusations of anthropomorphism. In 1646, a philosopher Rene Descartes contended that no nonhuman animal was sentient because the only evidence of sentience lay in the use of language to demonstrate an awareness of what things mean (Veit & Huebner 2020). Although the erroneous need for language to prove sentience crumbled in the face of affective neuroscience and cognitive ethology, focusing on an animal's affective experiences still raised concerns when trying to reconcile welfare with principles outlined in law and science-based policy (Beausoleil et al., 2023). The simple question posed by philosopher Thomas Nagel in 1974 when he wrote "What is it like to be a bat?" (p. 25) can be complex depending on a person's stance in relation to intensive animal agriculture, meat eating and animal welfare. To contemplate what—or even if—an animal is feeling during their short life and death can trigger the unbearable discomfort of cognitive dissonance, making Domain 5 a thought-provoking and contentious area to address.

As the Five Domains infiltrated animal welfare science, tentative hypotheses were formed for an animal's welfare state based on scientific indicators related to the animal's internal physical state, consistent behavioural and physiological measures and an understanding of nervous system pathways from sensory input to the generation of core mental experiences such as pleasure, pain and fear. An increasing understanding of the brain processing related to aversive and rewarding experiences and their manifestation as specific affects provided support to infer internal states, thus challenging accusations of anthropomorphism (Mellor, 2017). When integrated with biologist, ethologist, behavioural ecologist and writer Marc Bekoff's (1997) contention that being anthropomorphic does not ignore the animal's perspective, but rather makes this perspective more accessible using words that make sense from a human-centred perspective, Domain 5 becomes a valid measure of wellbeing.

WHY IS ANIMAL WELFARE IMPORTANT?

Despite consumers becoming increasingly concerned about the means by which animal products arrive on their plate—especially related to confinement, genetic manipulation and slaughter—in conjunction with animal advocacy movements gaining momentum, there is still a long way to go. The individual animal stories appearing in later chapters reveal that while many welfare gaps are closing, others still remain and are perpetuated by a range of factors. These include consumer demand, economic implications, the protective ignorance of cognitive dissonance, environmental and climate changes and cultural imperatives. It is the diversity of these root causes and the infinite and unexpected ways in which they combine to promote further suffering that prove so frustrating to those seeking to end the cycle of sentient animals living and dying as anonymous commodities.

The 2012 *Cambridge Declaration on Consciousness* made it clear that the capacity to experience pain and suffering does not solely belong to humans, and yet this simple fact has been disregarded in the lives of countless billions of animals and birds confined in industrial farm systems. The human experience of pain relies on two systems: a sensory-discriminatory system to convey information about the intensity and bodily location of the harmful stimulus to the somatosensory cortex of the brain, and a motivational-affective system that registers the unpleasantness of the stimulus and motivates adaptive action through the anterior cingulate cortex of the frontal lobe (DeGrazia, 2020). Humans are mammals, and so it is not surprising that other mammals also display the same basic neurological structure—the thalamocortical (thalamus-cortex) complex—in which pain processing occurs. Mammals have nociceptors, display protective behaviour towards injuries on their bodies and have an endogenous opioid system of natural opioids to regulate pain, emotion, stress responses and motivation. The evidence is indisputable—human and nonhuman animals are more alike than humans may care to accept, and yet animals continue to be used and abused for human desires. Anthrozoologist Hal Herzog's simple observation that "the way we think about other species often defies logic" (2010, p. 1) explains the complex psychological strategies that some people employ to avoid acknowledging that the plastic wrapped cuts of meat, blocks of cheese and cartons of milk came from a feeling "someone" rather than a "something".

Avian neuroanatomy differs from mammal neuroanatomy, but their brains are still wired to experience pain (DeGrazia, 2020). They may lack the cerebral cortex prominent in mammals, but the dorsal ventricular ridge seems to function in a similar role. They also have a cerebrum, or telencephalon, and somatomotor circuitry fully capable of generating the conscious experience of pain. And like mammals, they have nociceptive and endogenous opioid systems, demonstrate protective behaviours for injuries and learn to avoid noxious stimuli. In 2020, poultry meat represented 40 per cent of global meat production (Food and Agriculture Organisation of the United Nations, 2025). Despite the ongoing efforts of animal activists, many of these birds—fully capable of experiencing pain—were densely packed into unnatural environmental conditions, bred to grow so rapidly that their skeleton and internal organs could not hope to keep up and suffered immensely during their short lives.

Figure 1.2 Bird brains are wired differently than mammal brains, but they can still experience pain.

In 2021, Melanie Challenger, a researcher on the history of humanity and the natural world, commented: "The world is now dominated by an animal that doesn't think it's an animal" (Challenger, 2021, p. 1). This domination was sustained by the prefix in the word nonhuman, which created a "them and us" scenario and positioned animals as a separate and lesser category to human.

This dichotomy reinforces the tension between animal welfare and animal ethics. Intensive animal agriculture must treat farmed animals as commodities whose value lies in their utility or function as property, rather than as sentient beings. This means that their rights are determined by humans, the dominant species. Animal welfare has improved over the decades with the recognition that positive experiences, not just the avoidance of negative experiences, are fundamental to wellbeing. However, the perception persists that improving the experiences of farmed animals is linked to improving the quality of their output, a sentiment alluded to in the 2024 World Organisation for Animal Health [WOAH] vision paper, *Animal Welfare: A Vital Asset for a More Sustainable World* (WOAH, 2024). Acknowledging that a complete ban on farmed animal products would have long-term economic implications, the document instead advocates for the integration of animal welfare principles into animal production systems to avoid injury, ongoing fear and stress or brutal changes to their environment.

The welfare perspective adopts an anthropocentric perspective when adding that animals who are treated well are more productive, require less food and land and generate fewer pollutants, comments reminiscent of Temple Grandin's assertion that humane stunning at the slaughterhouse improves meat quality.

Animal ethics still has a long way to go to counter the many forces at work that prevent farmed animals from having a voice in an anthropocentric world. Farmed animal sanctuaries provide one means whereby a small number of animals are heard and respected, and in the following chapters, some of these animals from around the world give voice, albeit via a human voice, to what freedom and safety look like for them.

REFERENCES

Anthis, K. & Anthis, J.R. (2019). Global farmed and factory farmed animals estimates. *Sentience Institute*. https://www.sentienceinstitute.org/global-animal-farming-estimates

Beausoleil, N.J., Swanson, J.C., McKeegan, D.E.F., & Croney, C.C. (2023). Application of the Five Domains model to food chain management of animal welfare: Opportunities and constraints. *Frontiers in Animal Science, 4*. https://doi.org/10.3389/fanim.2023.1042733

Bekoff, M. (1997). Deep ethology. In M. Tobias & K. Solisti (Eds.), *Intimate relationships, embracing the natural world*. https://web-archive.southampton.ac.uk/cogprints.org/161/1/199710001.html

Birch, J., Burn, C., Schnell, A., Browning, H., & Crump, A. (2021). *Review of the evidence of sentience in Cephalopod Molluscs and Decapod Crustaceans*. London School of Economics and Social Science. https://www.lse.ac.uk/news/news-assets/pdfs/2021/sentience-in-cephalopod-molluscs-and-decapod-crustaceans-final-report-november-2021.pdf

Brambell, R.W.R. (1965). Report of the Technical Committee to enquire into the welfare of animals kept under intensive livestock husbandry systems. *Her Majesty's Stationery Office, London*. https://edepot.wur.nl/134379
Challenger, M. (2021). *How to be animal. A new history of what it means to be human*. Canongate Books Ltd.
Cochrane, A. (2014). Born in chains? The ethics of animal domestication. In L. Gruen (Ed.), *The ethics of captivity* (pp. 156–173). Oxford University Press.
Collinson, A. (2018). Overview of UK Animal Protection Legislation. *Michigan State University Animal Legal and Historical Centre*. https://www.animallaw.info/article/overview-uk-animal-protection-legislation
DeGrazia, D. (2020). Sentience and consciousness as bases for attributing interests and moral status: Considering the evidence and speculating slightly beyond. In L.S.M. Johnson, A. Fenton, & A. Shriver (Eds.), *Neuroethics and nonhuman animals* (pp. 119–133). Springer.
Edison, R. & Esvelt, K.M. (2020). On mitigating the cruelty of natural selection through human genome editing. In L.S.M. Johnson, A. Fenton, & A. Shriver (Eds.), *Neuroethics and nonhuman animals* (pp. 17–32). Springer.
Ferreira, V.H.B., Guesdon, V., & Calandreau, L. (2021). How can the research on chicken cognition improve chicken welfare: A perspective review. *World's Poultry Science Journal*, *77*(3), 679–698. https://doi.org/10.1080/00439339.2021.1924920
Festinger, L. (1957). *A theory of cognitive dissonance*. Stanford University Press.
Food and Agriculture Organisation of the United Nations (2025). Gateway to poultry production and products. https://www.fao.org/poultry-production-products/production/en/.
Grandin, T. (n.d.). Recommended stunning practices. *Dr Temple Grandin's Website*. https://www.grandin.com/humane/rec.slaughter.html
Grandin, T. (2022). Practical application of the Five Domains Animal Welfare Framework for supply food animal chain managers. *Animals*, *12*(20), 2831. https://doi.org/10.3390/ani12202831
Harrison, R. (1964). *Animal machines: The new factory farming industry*. Stuart.
Herzog, H. (2010). *Some we love, some we hate, some we eat*. Harper Collins Publications.
Jones, (2020). Speciesism and human supremacy in animal neuroscience. In L.S.M. Johnson, A. Fenton, & A. Shriver (Eds.), *Neuroethics and nonhuman animals* (pp. 99–115). Springer.
Joy, M. (2009). *Why we love dogs, eat pigs, and wear cows*. Conari Press.
Ladak, A. & Anthis, J.R. (2022). Animals, Food, and Technology (AFT) Survey: 2021 update. *Sentience Institute*. https://www.sentienceinstitute.org/aft-survey-2021
Levy, D.M. 1944. On the problem of movement restraint: Tics, stereotyped movements, hyperactivity. *American Journal of Orthopsychiatry*, *14*, 644–671.
Littlewood, K.E., Heslop, M.V., & Cobb, M.L. (2023). The agency domain and behavioral interactions: Assessing positive animal welfare using the Five Domains Model. *Frontiers in Veterinay Science*, *10*. https://doi.org/10.3389/fvets.2023.1284869

Mellor, D.J. (2017). Operational details of the Five Domains Model and its key applications to the assessment and management of animal welfare. *Animals, 7*(60). https://doi.org/10.3390/ani7080060
Mellor, D.J., Beausoleil, N.J., Littlewood, K.E., McLean, A.N., McGreevy, P.D., Jones, B., & Wilkins, C. (2020). The 2020 Five Domains Model: Including human–animal interactions in assessments of animal welfare. *Animals, 10*(10), 1870. https://doi.org/10.3390/ani10101870
Nagel, T. (1974). What is it like to be a bat? *Philosophical Review, 83*, 435–450.
orzechowski, k. (2024). *Global animal slaughter statistics & charts.* Faunalytics. https://faunalytics.org/global-animal-slaughter-statistics-and-charts/
Piazza, J., Matthew, B.R., Loughnan, S., Luong, M., Kulik, J., Watkins, H.M., & Seigerman, M. (2015). Rationalizing meat consumption. The 4Ns. *Appetite, 91*, 114–128.
Protection of Animals Act 1911. https://www.legislation.gov.uk/ukpga/Geo5/1-2/27/data.pdf
Regan, T. (1983). *The case for animal rights.* University of California Press.
Robertson, I. & Goldsworthy, D. (2022). Recognising and defining animal sentience in legislation: A framework for importing positive animal welfare through the Five Domains Model. *Monash University Law Review, 48*(1), 244–271.
Safina, C. (2015). *Beyond words: What animals think and feel.* Henry Holt and Company, LLC.
Veit, W. & Huebner, B. (2020). Drawing the boundaries of animal sentience. *Animal Sentience, 29*(13). https://doi.org/10.51291/2377-7478.1595
Voiceless (2015). *The animal law toolkit* (2nd ed.). https://voiceless.org.au/wp-content/uploads/2015/12/Voiceless_Animal_Law_Toolkit_2ND_Edition.pdf
Wacks, R. (2024). *Animal lives matter. The continuing quest for justice.* Routledge.
Webster, J. (2007). *New trends in animal welfare.* ISAH: Tartu June 2007. https://www.isah-soc.org/userfiles/downloads/proceedings/Plenary_2007/JohnWEBSTER.pdf
WOAH (2024). *Animal welfare: A vital asset for a more sustainable world.* Paris, 8 pp. https://doi.org/10.20506/woah.3440. Licence: CC BY-SA 3.0 IGO.

2

Farmed animal sanctuaries: A growing movement

For millennia, humans have attempted to reshape the world and its inhabitants to suit their own evolving needs. As the self-appointed dominant species, they wreaked havoc, leaving each generation to bear the legacy of the previous generations' actions. Many animals have experienced the human need to control with results ranging from domestication to extinction, depending on the complex relationship they shared with humanity. The implicit—and at times, explicit—reluctance to accept that animals do not simply exist to be managed, protected, utilised or abused by humanity blocked any perceptions of them as unique individuals with feelings. These feelings may not manifest in a human-like manner, but they were important nonetheless. Even those animals hidden from urban dwellers within industrial farm systems and destined for an early death had capacity to feel in ways that those who ate or used their bodies seemed unable to acknowledge without risk of cognitive dissonance. However, the true extent of a farmed animal's capacity to feel becomes undeniable when observing them living out their lifespan in the safety of a farmed animal sanctuary.

DOMESTICATION

The domestication of animals changed the evolutionary trajectory of humanity and consolidated a long-standing belief that animals existed for human benefit. Long ago, Greek philosopher Aristotle (384-322 BCE/before common era) asserted that: "...plants are evidently for the sake of animals, and animals for the sake of Man; thus Nature, which does nothing in vain, has made all things for the sake of Man" (Clutton-Brock, 2012, p. 1), a sentiment that lingers into the twenty-first century.

Animals held within a human community that maintains complete control over their breeding, territory and food supply are considered domesticated rather than simply tamed (Clutton-Brock, 2012). Domestication requires a combination of biological and cultural processes, starting with the separation and taming of a few animals from a wild species. Once they breed and produce viable offspring,

DOI: 10.1201/9781003518853-3

these form a unique group that continues to interbreed and increase over generations. However, where small genetic changes through natural selection to adapt to their humanised environment were supplemented by artificial selection for specific traits that the human community deemed valuable, the group became genetically separate from the founding species. This combination of human manipulation for desired traits and their own adaptation to new environments eventually left domesticated animals dependent on their human caretakers who simultaneously became dependent on the animal for food, animal products or labour.

Domestication of animals and the agricultural revolution, which allowed humans to move from hunting and gathering subsistence to settling in communities, has been described as both humanity's freedom from the constraints of nature and the catalyst for a global ecological crisis (Clutton-Brock, 2012). The journey from a species' first association with humans to their final exploitation through intensive animal agriculture may vary for each animal, but the end result would be forced separation from their culture of origin and adaptation to a human-controlled way of life. Somewhere along the way perceptions of sentience and emotional capacity in commercially farmed animals disappeared, while some companion animals were elevated to family member status. The irony of this demarcation is clear in anthropologist Hal Herzog's (2010) aptly titled book *Some We Love, Some We Hate, Some We Eat: Why It's so Hard to Think Straight About Animals.*

EMOTION AND DOMESTICATION

Even before the 2012 *Cambridge Declaration on Consciousness*, animal sentience was being discussed and debated in academic circles. Professor of zoology Donald Griffin wrote in 1981: "The belief that mental experiences are a unique attribute of a single species is not only unparsimonious; it is conceited" (p. 170), a perspective that continued to be validated in scientifically rigorous ways through the integration of neuroscience, evolutionary biology, comparative psychology, animal welfare science and philosophy. When the *Cambridge Declaration on Consciousness* formalised scientific consensus that humans were not the only conscious beings, it seemed a logical assumption that the moral failure of farmed animal abuse could no longer be denied. If these animals had neural substrates capable of supporting conscious experiences and awareness of their surroundings, surely they could no longer be treated as property and resources. Cruelty-free consumption and global veganism should have been a logical and inevitable outcome.

More than a decade passed before the 2024, *New York Declaration on Animal Consciousness* was released, again underscoring the ability of animals to suffer and experience wellbeing (Andrews et al., 2024). Based on strong scientific support that animals and birds are conscious, the declaration contended that humans have an ethical obligation to consider the welfare risks and interests of those animals as individuals. Recognising animals as individuals with unique emotional and behavioural dispositions is fundamental to farmed animal sanctuaries, where anthropocentric perceptions of animals as resources and commodities are firmly rejected.

While sentience is an animal's capacity to experience and respond to both positive and negative sensations, emotion generates feelings about the sensations being experienced. In 2006, philosopher Martha Nussbaum (2006) included emotion as an entitlement when proposing a guideline for animal welfare. Item 5 of the entitlements stated:

> 5. Emotions. Animals have a wide range of emotions [...] they are entitled to lives in which it is open to them to have attachments to others, to love and care for others and not to have those attachments warped by enforced isolation or the deliberate infliction of fear.
>
> *(p. 397)*

When examining emotion among animals—including human animals—researchers Elizabeth Paul and Michael Mendl (2018) identified some common features that positioned emotion as a subjective, physiological, behavioural and cognitive response to a stimulus or event of importance to the individual. Emotion could be valenced (either pleasant or unpleasant) and varied in activation and duration from fleeting to days. The type of emotion depended on the emotion-provoking external or internal event, as well as the importance of that event to the individual's goals and wellbeing.

Neuroscientist and psychobiologist Jaak Panksepp (2004) introduced the term affective neuroscience in the 1990s when exploring cross-species neuroscience. Modern affective neuroscience has identified some specific brain structures involved in emotions, including core systems of the peri-aqueductal gray matter in the upper brainstem, the hypothalamus and amygdala and the anterior cingulate gyrus and nucleus accumbens (Turnbull & Bar, 2020). An important observation arising from Panksepp and other researchers' work was that the subjective feeling component of emotion arose from ancient subcortical structures common to humans, mammals and birds and did not require the involvement of the neocortex. This removed emotional capacity from perceived intelligence and negated arguments that the "stupidity" of some animals rationalised their commodification for humanity. However, the brain can also show structural and functional plasticity in response to stressful situations and prolonged suffering. In humans, changes to neural circuitry can affect cognition, decision-making, anxiety and mood (McEwen, 2017), raising questions about the neurological damage incurred by keeping farmed animals densely packed into unnatural environmental conditions.

With mounting scientific evidence of sentience and emotion in farmed animals, there should be no justification to continue confining, transporting and slaughtering billions of animals to meet human demand. However, for consumers to acknowledge this troubling fact is to invite guilt and the mental anguish of cognitive dissonance.

INDUSTRIAL FARM SYSTEMS

Despite advancements in perceptions and assessments of animal welfare over the decades, industrially farmed animals still remain property with commodified bodies for human use and abuse. Genetic modifications, disruption of family ties and the overuse of pharmaceuticals—all in the name of productivity—have negatively affected the lives of billions of sentient beings. For many animals, it is too late to undo what has been done, and they will live and die in confinement during their human-determined lifespan. Despite the rapidly expanding farmed animal sanctuary movement, around three-quarters of farm animals globally continue to live in factory farm settings (Anthis & Anthis, 2019).

While it is difficult to identify a specific definition of factory farms and industrial farm systems, they are commonly described as "concentrated animal feeding operations" where animals are kept and raised in confined conditions. Specifically, "animals have been, are, or will be stabled or confined and fed or maintained for a total of 45 days or more in any 12-month period" in an area that does not produce vegetation (United States Environmental Protection Authority, 2024).

Human population growth coinciding with increased productivity of the Industrial Revolution in the eighteenth century meant it was only a matter of time before industrialisation reached the animal agriculture sector. Advances in selective breeding allowed manipulation of animals at the genetic level to more efficiently meet ever-growing demand for food and animal products. Artificial insemination techniques streamlined breeding, antibiotics reduced outbreaks of disease in dense concentrations of animals, assembly lines for egg collection and milking machines that could milk hundreds of cows per hour—all contributed to turning domesticated animals into commodities to be manipulated, used and abused for human benefit. Many smaller farms diminished, unable to compete economically with the concentrated "mega factory farms" that housed thousands of animals, with devastating consequences for the animals and the environment. As animal agriculture became one of the leading drivers of the climate crisis, a World Bank Report confirmed that food and agriculture generated almost a third of global greenhouse emissions, 60 per cent of which was attributable to meat and dairy farming (Sutton et al., 2024). However, by the twenty-first century, factory farming had also become a global necessity to service the high demand for animal products.

The earlier decades of animal advocacy continued with escalating activism from animal welfare groups, environmentalists and sectors of the general population. In 1993, a new tactic to combat factory farms gained momentum in Australia before spreading globally (Scott-Reid, 2023). After receiving footage of atrocities at an egg farm, Australian animal activist Patty Mark went on to the property to document what she saw and offer aid to any distressed animals. The animal status of property meant removing them without the owner's permission was theft. There was a loophole though—if the animal was injured, ill or otherwise suffering, that could change its status and allow the activists the legal right to save an animal in need of help. This rationale and tactic became known as "open rescue", putting factory farms in the spotlight and inspiring a wave of grassroots activists.

Social media provided a global vehicle to highlight the plight of these sentient beings. Horrific images of confinement methods such as gestation crates for pigs, veal crates for calves and cages for egg-laying chickens relied on "moral shock" to motivate public outrage against animal cruelty activities hidden behind closed doors. However, there was also the risk of a counter-response when people opted to turn away from distressing images of tiny male chicks about to be macerated, grief-stricken cows deprived of baby after baby to ensure their milk flowed, tiny goats born at the slaughter yard and flung to one side as they awaited death, pregnant pigs trapped in metal crates no bigger than their bodies or bloodied deformed laying chickens crammed into barren cages. The risk of cognitive dissonance created by these images had the potential to reinforce a refusal to connect eating meat with the life of the living animal. Occasionally, the publicity surrounding animal abuse and death on an unimaginable scale became unavoidable during events such as live animal exports caught up in maritime disasters or mass "depopulating" during infectious disease outbreaks. However, when it became too difficult for people to remain ignorant of the darker side of commercial agriculture, another powerful psychological strategy was available to reinforce a purposeful dissociation from moral blame.

Psychophysical or psychic numbing serves as a flawed protective mechanism by instilling a sense of indifference and removal of personal responsibility when events become too overwhelming to contemplate or remedy. Psychologist Paul Slovic (2007) proposed an antidote to the inertia associated with psychophysical numbing: in order to elicit compassion, there needs to be an identified victim with a face and a name. Slovic described a situation that arose during the 2001 outbreak of foot and mouth disease and the subsequent slaughter of millions of cattle in the United Kingdom. When the disease disappeared and the killings continued as a precaution against future infection, appeals from animal rights groups to cease this slaughter went unheard until the media published the image of a single calf. Phoenix, a cute 12-day-old calf found huddled next to her dead mother amidst a pile of carcasses that had been culled "just in case", was uncontaminated with foot-and-mouth, but still destined for slaughter. Phoenix achieved what death statistics, facts and activism could not—she was granted a reprieve from the national slaughter policy and provided a catalyst for the United Kingdom government to rework its approach to indiscriminate mandatory slaughter.

This "singularity effect" embodied by Phoenix allowed an understanding of the scale of an atrocity to emanate from an individual. Mother Teresa (1910–1997), the Catholic nun who dedicated her life to saving the poor and destitute, succinctly summed up this effect in a famous quote: "If I look at the mass I will never act. If I look at one, I will" (Slovic, 2007, p. 80).

Farmed animal sanctuaries allow animal residents to become "the one" as they move from the anonymity of industrial farm systems to a life where they have a name, a story and a future. Their images and stories have capacity to stimulate empathy and motivation for change within the wider public without inducing a sense of helplessness at the magnitude of the countless animals who could not be saved.

FARMED ANIMAL SANCTUARIES

Farmed animal sanctuaries have emerged around the world, sharing a common mission to rescue, rehabilitate and care for animals who were once commodities in an exploitive human-controlled environment (Rosenfeld, 2024). There may be individual variations between sanctuaries depending on location and resources, but in general, they ensure that animals have security—often in a multi-species environment—for the rest of their natural life.

Many founders cite one pivotal moment or animal that provided the catalyst to move from being an animal lover and/or animal activist to rescuing, rehabilitating and providing a safe home for as many individuals as possible. In Australia, Edgar's Mission owed its genesis to a pig named Edgar Alan Pig, a British actor, James Cromwell (aka Farmer Hoggett) from the hit movie *Babe* and the tenacity of the sanctuary's founder, Pam Ahern, to highlight the flawed animal protection legislation that excluded farmed animals (Edgar's Mission, 2025).

Saint Beans Animal Sanctuary for Lambs began after a tiny blob of white bounded up to a man on an icy morning in New Zealand. Abandoned as a newborn and at risk of hypothermia, Beans would instead become the first resident of a sanctuary dedicated to sheep and lambs.

Peruse the website of Farm Sanctuary in the USA, and the story of Hilda, the sheep rescued from a pile of dead animals, demonstrates the power of the singularity effect to create change for future rescued farmed animals (Farm Sanctuary, 2025). The story of Esther, the social media wonder pig, is immortalised in both a book and the Happily Ever Esther Farm Sanctuary in the USA, which was established to accommodate the supposed mini pig who grew into a 272 kilogram (600 pound) full-sized charismatic dynamo (Jenkins & Walter, 2016). The Donkey Sanctuary started with a donkey named Naughty Face and the founder Dr Svendsen's growing awareness of the appalling state of donkeys at a donkey market. By 1973, the Donkey Sanctuary was a registered charity that went on to become a global force to be reckoned with in the fight for donkey and mule welfare rights (The Donkey Sanctuary, 2025).

In 2017, Voiceless, an Australian-based organisation seeking an equitable world where animals can flourish, published their first "Voiceless Animal Cruelty Index". This index tracked the animal welfare performance across 50 countries identified as the largest producers of farmed animal products in the world. The updated index released in 2020 provided some disturbing statistics:

> According to the latest FAO statistics, 77 billion farmed animals are slaughtered each year for human consumption globally. This equates to 211 million animals per day, 8.8 million per hour and 2,442 animals per second. The vast majority are factory farmed,

> where they are confined for the entirety of their lives in sheds or cages, subjected to unnatural diets and environments, and exposed to painful practices and procedures, often without pain relief.
>
> *(Voiceless, 2020)*

With most farmed animal sanctuaries unable to offer refuge to more than a few hundred residents at a time, there will always be animals who remain trapped within industrial farm systems for the entirety of their lives. The singularity effect, so evident in the story of Phoenix the calf, is enacted every day among the

Figure 2.1 The entrance to Edgar's Mission, where a donkey grazes peacefully by the sign. (Photograph taken by the author at Edgar's Mission, Lancefield, Victoria, Australia, and reproduced with their permission.)

Figure 2.2 Young Beans, a lamb found abandoned as a newborn. (Image supplied by Saint Beans Animal Sanctuary for Lambs and reproduced with their permission.)

animals who make it to a sanctuary and become ambassadors and advocates for the billions of farmed animals who will never leave the industrial farm systems.

Establishing and maintaining a sustainable and ethical farmed animal sanctuary can be rewarding—but personally and financially draining when seeking to address the physical, behavioural and social needs of a diverse range of animals within the constraints of spatial and resource limitations. Many of the animals who make it to a farmed animal sanctuary carry the legacy of human manipulations and abusive treatment. They require months and even years to recreate themselves following traumas deeply embedded in their psyche and compounded by any physical disabilities inflicted on their bodies through domestication and abuse. Funding to physically, emotionally and medically support these animals can be reliant on grants and public donations, which can be fickle, especially in times of economic hardship.

This renders many sanctuaries reliant on dedicated founders, family members, friends and volunteers, to meet the physical, emotional and medical needs of residents as well as property maintenance. As with many rescue organisations, the relentless demand to take in more and more animals can dominate the needs of the humans seeking to help them. Rejecting requests to rescue an animal in need when finances and physical resources are depleted can bring feelings of guilt in already exhausted human carers. Compassion fatigue, described as the emotional and physical toll of caring for others, is common among people who work in animal welfare as well as those who work with trauma survivors. Working with traumatised animals can thus leave farmed animal sanctuary workers vulnerable in both domains.

RESCUE, EDUCATION OR ADVOCACY?

Urban dwelling consumers, industrial farm systems and slaughterhouses are separated geographically, visually and cognitively. As animals move from factory farm to slaughterhouse, both operations strive to limit public access to the realities of their day-to-day activities. Farmed animal sanctuaries provide a crack in this secrecy through which those who choose to visit are confronted with the living, sentient producers of the meat, eggs, dairy and yarn that are perceived as indispensable in contemporary society. Connecting the unique characters who populate these sanctuaries with the faceless billions of farmed animals triggers cognitive dissonance as a preliminary step in the sanctuary's educational role and the process of behavioural change. Whether the change takes hold immediately, at a later stage or at all can be difficult to gauge.

Human change behaviour has been conceptualised as occurring in six stages (Prochaska & Norcross, 2014). Each stage comprises specific groupings of attitudes, intentions and behaviours that must be completed across time to enact movement to the next stage of change. These stages of pre-contemplation, contemplation, preparation, action, maintenance and termination when change is complete may never be fulfilled if a person chooses alternative strategies to counter any cognitive dissonance after meeting individual residents. Strong social norms and pressure, pre-existing beliefs or an unwillingness to give up something perceived as pleasurable can negate any attitude-behaviour inconsistency upon returning to an urban lifestyle. However, many visitors and supporters of farmed animal sanctuaries have already chosen not to consume animal products, and each visit becomes a reaffirmation of the need to protect these animals.

Farmed animal sanctuaries have a responsibility to balance the animal residents' freedom from captivity and human control in their new home against concerns for their safety and wellbeing (Abrell, 2020). Boundaries can blur between the need to minimise captivity as much as possible and the practicalities of animal care and control within the designated sanctuary setting. Captivity without control can be difficult to achieve, as philosopher Lori Gruen pointed out when defining the concept of captivity as: "a condition in which a being is confined and controlled and is reliant on those in control to satisfy her basic needs" (2011, p. 133).

This descriptor is subtly reinforced in the language used by the sanctuaries themselves when making it clear that animals are "as free as possible" (Jones, 2014, p. 91). There are fences to keep animals within the confines of the sanctuary and away from human complaints, strict routines to meet daily demands and limited workforce power and certain medical procedures such as sterilisation to comply with sanctuary rules, all of which reinforce the animals' lack of complete autonomy as the price for safety. Miriam Jones (2014), co-founder of VINE Sanctuary in the USA, commented that at a micro-agency level, sanctuaries respect animals as individuals with day-to-day preferences for food, sleeping and gathering spots, activities and friends, but choices at the macro-agency level largely reside in the hands of the sanctuary human carers.

Two philosophical paradigms have traditionally influenced animal advocacy within the farmed animal sanctuary movement: welfarism and animal rights (Abrell, 2020). The welfarism approach seeks to improve welfare and reduce suffering. These factors also contribute to the rights approach objective to end harmful treatment and the cause of suffering by addressing the property status of animals and granting them certain legal rights. Writers and philosophers Sue Donaldson and Will Kymlicka (2015) describe a "refuge+advocacy model" that promotes safety from harm, neglect, exploitation or commodification parallel with the provision of basic needs, awareness and advocacy. These sanctuaries may resemble a traditional family farm in terms of location, buildings, selection and grouping of animals, and as the animals and their needs grow, the sanctuary seeks not-for-profit status to raise funds.

Many sanctuaries also operate education programmes of some sort as part of their advocacy role. The animals are positioned as ambassadors whose "happy ending" (in life or death) provides proof that the wrongs of factory farming are survivable for the lucky few in the refuge of a farmed animal sanctuary. Many remain crippled, terminally ill or with shortened lifespans from intensive breeding and production practices, but treatment of these infirmities with medications, prosthetics, physiotherapy and socialisation further exemplifies the horrors of factory farming and consolidates farmed animal sanctuaries as providers of compassionate care. However, instrumentalising animal ambassadors as advocates and fund raisers for the animal protection movement requires careful monitoring so as not to encroach on an animal's autonomy in regulating its proximity to unfamiliar humans (Abrell, 2020).

Recognising the need to create change in a new generation, some sanctuaries offer targeted programmes for children. For example, at Flow Farm and Sanctuary in China, children explore and learn in an environment that fosters empathy and compassion for animals and the natural world. Deeply engrained adult consumption habits can prove problematic to change, but young children lack an extended history of meat consumption and a repertoire of strategies to counter cognitive dissonance. Targeting children through age-appropriate activities and interactions at sanctuaries provides opportunity for change before consumption patterns become entrenched. While still linked to the economic role that animal ambassadors play in raising funds to support their up-keep, this targeted educational component takes on a future-focused orientation to enact long-term behaviour change.

Animal Charity Evaluators, an organisation that conducts charity evaluations to identify organisations likely to make the most difference for animals through public donations, cited education and advocacy as having a greater long-term impact than the number of animals saved by farmed animal sanctuaries each year (Bockman, 2015). Acknowledging that the insatiable need for funding in the direct care of animals could detract from focusing on education and outreach, their 2024 *Farmed Animals Guide* instead recommended charities that focused on societal change such as corporate and institutional outreach and legal advocacy (Animal Charity Evaluators, 2024).

When exploring the challenges of farmed animal sanctuaries, Donaldson and Kymlicka contrasted the "refuge+advocacy model" with an "intentional community model" committed to creating communities that were "more spacious, complex, varied, open, unpredictable, and free, in which animals are actively enabled to have a say in how they will live" (p. 63). As part of an intentional community, they proposed that animals would be in a better position to identify what they wanted and how they wanted to live, which may be different to what humans anticipated. Different possible lives within this model could allow for interspecies friendships when segregation was not based on human assumptions or convenience. The authors acknowledged the logistical challenges and inherent risks associated with free association (for example, animal size, gender, conflict and environmental needs), but they contended that sharing control with animals should come first, followed by creative solutions and risk mitigation. Redefining farmed animal sanctuaries as intentional communities rather than just safe havens for "refugees in need of humanitarian care" was a necessary step to progress animal advocacy into the future, they contended, adding: "If animals could be seen as pioneers of a just future, rather than as ambassadors of an unjust present, this would surely have a more transformative effect on the humans who encounter them as co-citizens of FASes [farmed animal sanctuaries]" (p. 68).

ETHICAL IMPLICATIONS

There is no denying that farmed animal sanctuaries play a crucial role in the animal welfare message. They highlight the individuality of farmed animals as sentient beings with a past, present and future. They may not be a perfect solution to global animal exploitation, but there is no one-size-fits-all remedy to the way humans have impacted the lives of animals for centuries. Whether they are "happy" living on farmed animal sanctuaries is something that can only be inferred. However, when contrasted with the environments from which most of them originated, a logical assumption would err on the side of improvements in their physical and psychological circumstances.

Animals need rescuing because some humans have chosen to hurt and kill them in the name of global demand for food. Most cannot be rewilded because they remain trapped in unsustainable bodies engineered by humans to increase supply. Some like the broiler chicken whose legs have shattered under the unnatural weight die quickly because of these manipulations, while others, such as the calf born to a dairy cow pregnant at time of rescue, can remain in safety for the whole of their natural life. Sanctuaries may contribute to the emotional and physical repair of these animals, but as philosopher and animal ethics writer Karen Emmerman (2014) points out, they can in no way provide restitution for the harms inflicted on them prior to rescue. Instead, as each animal ambassador dies and is grieved as a unique individual, the message remains the same. Only the abolition of inhumane intensive farm systems could make up for the wrongs inflicted on these sentient beings in a human-dominated world.

In the following chapters, the stories of animals living in 23 farmed animal sanctuaries worldwide provide a glimpse into the unique lives of a range of animals who were never meant to live beyond their human-designated use-by date. Their diverse personalities, preferences and emotions are a reminder of Marc Bekoff's (2005) contention that: "the more we study animals and the more we learn about 'them' and 'us', the more we discover there is no real dichotomy or nonnegotiable gap between animals and humans, because humans are, of course, animals" (p. 12).

REFERENCES

Abrell, E. (2020). Animal sanctuaries. In B. Fischer (Ed.), *The Routledge handbook of animal ethics* (pp. 569–577). Routledge.

Andrews, K., Birch, J., Sebo, J., & Sims, T. (2024). *Background to the New York declaration on animal consciousness.* https://sites.google.com/nyu.edu/nydeclaration/declaration.

Animal Charity Evaluators (2024). *Farmed animals guide.* Guide to Help Farmed Animals.

Anthis, K. & Anthis, J.R. (2019). Global farmed and factory farmed animals estimates. *Sentience Institute.* https://www.sentienceinstitute.org/global-animal-farming-estimates

Bekoff, M. (2005). *Animal passions and beastly virtues: Reflections on redecorating nature.* Temple University Press.

Bockman, J. (2015). The value of sanctuaries, and how to maximize their impact. *Animal Charity Evaluators.* https://animalcharityevaluators.org/blog/the-value-of-sanctuaries-and-how-to-maximize-their-impact/

Clutton-Brock, J. (2012). *Animals as domesticates: A world view through history.* Michigan State University Press.

Donaldson, S. & Kymlicka, W. (2015). Farmed animal sanctuaries: The heart of the movement? A socio-political perspective. *Politics and Animals, 1*(1), pp. 50–74.

Edgar's Mission (2025). *About us.* https://edgarsmission.org.au/sanctuary/about/

Emmerman, K. (2014). Sanctuary, not remedy. In L. Gruen (Ed.), *The ethics of captivity* (pp. 213–230). Oxford University Press.

Farm Sanctuary (2025). *Rescue and refuge.* https://www.farmsanctuary.org/rescue/

Griffin, D. (1981). *The question of animal awareness: Evolutionary continuity of mental experience.* (Revised and enlarged edition). William Kaufmann, Inc.

Gruen, L. (2011). *Ethics and animals: An introduction.* Cambridge University Press.

Jenkins, S. & Walter, D. (2016). *Esther the Wonder Pig. Changing the world one heart at a time.* Affirm Press.

Jones, M. (2014). Captivity in the context of a sanctuary for formerly farmed animals. In L. Gruen (Ed.), *The ethics of captivity* (pp. 90–101). Oxford University Press.

McEwen, B.S. (2017). Neurobiological and systemic effects of chronic stress. *Chronic Stress, 1*, February 2017. https://doi.org/10.1177/2470547017692328
Nussbaum, M.C. (2006). *Frontiers of justice: Disability, nationality, species membership.* Harvard University Press. https://doi.org/10.4159/9780674041578
Panksepp, J. (2004). *Affective neuroscience: The foundations of human and animal emotions.* Oxford University Press.
Paul, E.S. & Mendl, M.T. (2018). Animal emotion: Descriptive and prescriptive definitions and their implications for a comparative perspective. *Applied Animal Behaviour Science, 205*, 202–209. https://doi.org/10.1016/j.applanim.2018.01.008
Prochaska, J.O. & Norcross, J.C. (2014). *Systems of psychotherapy: A transtheoretical analysis* (8th ed.). Cengage Learning.
Rosenfeld, H. (2024). Deviant accumulation at farmed animal sanctuaries. *EPE: Nature and Space, 7*(1), 271–287. https://doi.org/10.1177/25148486231167870
Scott-Reid, J. (2023). The Open Rescue Movement for farm animals, explained. *Sentient Media.* https://sentientmedia.org/open-rescue-movement/
Slovic, P. (2007). 'If I look at the mass I will never act': Psychic numbing and genocide. *Judgment and Decision Making, 2*, 79–95. https://ssrn.com/abstract=1569098
Sutton, W.R., Lotsch, A., & Prasann, A. (2024). "Recipe for a liveable planet: Achieving net zero emissions in the agrifood system". *Agriculture and Food Series. Overview booklet.* World Bank, Washington, DC. License: Creative Commons Attribution CC BY 3.0 IGO. https://openknowledge.worldbank.org/server/api/core/bitstreams/a0431f46-c65e-441a-9eed-8783d8b00272/content; https://sentientmedia.org/un-green-climate-fund-factory-farming/
The Donkey Sanctuary (2025). *Our history.* https://www.thedonkeysanctuary.org.uk/what-we-do/who-we-are/our-history
Turnbull, O.H. & Bar, A. (2020). Animal minds: The case for emotion, based on neuroscience. *Neuropsychoanalysis, 22*(1–2), 109–128. https://doi.org/10.1080/15294145.2020.1848611
United States Environmental Protection Authority (2024). *Animal Feeding Operations (AFOs).* https://www.epa.gov/npdes/animal-feeding-operations-afos
Voiceless (2020). *Voiceless Animal Cruelty Index 2020.* https://vaci.voiceless.org.au/about-the-vaci/

3

Cattle: Sentient animals, not "raw materials"

INTRODUCTION

The dairy industry keeps some deadly secrets, according to the Australian animal rights organisation, Farm Transparency Project (2025). They cite instances of newborn calves crying out in distress as they are dragged from their equally distressed mothers, and the fate of discarded bobby calves who are slaughtered within hours or days of birth. Newborn dairy calves who are not with their mothers are called "bobby calves". They are deemed surplus to dairy industry requirements when they are either not suitable (if male) or not required (if female) for the milking herd. Their mothers remain in the dairy herd until a four- to seven-year relentless cycle of pregnancy, birth and separation leaves them "spent", and they are sent to slaughter for being unable to produce sufficient milk to justify their existence.

The beef cattle industry also has secrets, including cattle with a natural lifespan of up to 25 years being slaughtered at 18 months or less. During those 18 months, they may be kept in cramped fenced areas called feedlots, subjected to painful procedures, including castration, branding and disbudding with a hot iron, or packed onto a crowded boat for live export (Farm Transparency Project, 2022). Much of this happens away from the public eye, only surfacing through the persistent work of animal activist and advocacy groups.

However, on a winter's evening in July 2024, residents in an inner-suburban Melbourne suburb in Australia were witness to the death of 22 sentient animals when a truck carrying dairy steers and calves to slaughter crashed into an overhead bridge. The top sections of the truck were crushed, leaving cattle with excruciating injuries, dying or dead. Others jumped from the truck and ran down suburban streets. As volunteers from Vets for Compassion sought to care for the injured, the terrified escapee animals were loaded onto another truck to continue their journey to slaughter (Vets for Compassion, 2024). What made this crash different—and harder to ignore—compared to other truck rollovers when carrying

DOI: 10.1201/9781003518853-4

farmed animals was the location. This happened in full view of urban dwellers in the heart of a capital city, not in a remote sparsely populated rural area. Eight hours of road closures and train line disruptions made it hard to ignore the inconvenient (to humans) accident that claimed the lives of so many animals.

As debate raged around fault and blame for the accident, to a lesser extent, questions were raised as to whether the surviving traumatised cattle should be given a reprieve and sent to live out their days on a farmed animal sanctuary. Dairy steers are considered property to be disposed of at will rather than sentient beings deserving of compassion, and so they continued their journey to meet another type of death at the slaughterhouse. When asked if cattle had an awareness of death, Dr Temple Grandin (2001) responded that cattle at the slaughter yard and cattle in a feedlot veterinary chute showed no discernible behavioural differences, suggesting that they had no understanding of death. However, having experienced the visual, auditory and olfactory stimuli of death at the accident site, these traumatised survivors may already have had a heightened awareness of death when identifying similar sensory stimuli at the slaughter yard, thus making their final hours of life even more distressing.

Whether any animal has an awareness that death is approaching may be unclear, but some appear to recognise and try to break free from the aversive sights, sounds and smells of slaughter yards. In the early evening of a summer's day in 2021, 40 cows escaped from a slaughterhouse in Los Angeles, USA. As with the Melbourne incident, the hidden life and death of farmed cows became a public spectacle as these terrified animals ran for their lives. In their desperate search for safety, fences were shattered, cars damaged and gardens and bushes flattened as the Los Angeles County Sheriff's Department and ranchers tried to stop these terrified, sensitive and bleeding animals (Farm Sanctuary, 2021). One cow was shot dead, two made it to a farmed animal sanctuary and the rest resumed their inevitable journey to slaughter. Images of these animals fighting for their survival jolted some urban dwelling consumers into the debate about the ethics of eating animals, just as the scenes of carnage in the Melbourne truck accident would do three years later.

While the sheer number of escapees in the Los Angeles incident was unusual, many single cows have managed to flee certain death with less publicity. Whether cows can think about their future has no empirical evidence, but anecdotal reports suggest that this is a very real possibility, adding a new dimension to these escapee reports. If this is found to be the case—especially as a driving force to flee death at the slaughterhouse—this negates any value in the concept of humane stunning and slaughter when they know it is coming. It also raises serious questions about the implications of eating these sentient beings who were domesticated over 9,000 years ago and selectively bred for human consumption.

SOME BACKGROUND HISTORY

As with many domesticated farm animals, humans first hunted wild cattle, called aurochs or *Bos primigenius*, before domesticating them to accommodate a lifestyle change to agriculture (Clutton-Brock, 2012). Wild cattle of the Stone Age and New Stone Age (Neolithic) periods stood up to two metres (6.5 feet) at the shoulder and weighed more than one tonne (2,200 pounds). Exploitation for their meat, horns, hides and any other useful body parts saw them hunted to extinction, to be replaced by the domestic cattle used for labour, meat, milk, leather and manure.

Described as a "walking larder" (Clutton-Brock, 2012, p. 42), cattle were at the forefront of the secondary products revolution with milk production. Archaeological findings place milk production in northeastern Anatolia (Asia Minor) by the seventh millennium BCE (before common era), in eastern Europe by the sixth millennium BCE and in Britain by the fourth millennium BCE. Traditionally, women did most of the milking by hand, but by the 1950s, milking machines had changed farm dynamics in many developed nations to allow larger herds as a precursor to industrialised farm settings.

Globally, there are more than 1,000 breeds of domestic cattle, which are largely divided into two types: those without a hump (*Bos taurus*) and those with a hump (zebu cattle, *Bos indicus*) (Lidfors, 2022). Insight into whether both descended from the one species of aurochs or humped cattle came from a separate species widespread in India consolidated with the refinement of molecular analysis capabilities (Clutton-Brock, 2012). The genetic distance between humped and non-humped cattle seemed too great to have one ancestor, and the investigation shifted to differentiating the two ancestors. Evidence pointed towards non-humped cattle originating from the *Bos primigenius* and humped cattle from a group of aurochs called *Bos namadicus* within three primary centres of domestication: western Asia, North Africa and India.

Non-humped cattle, including the common dairy breeds of Holstein, Ayrshire and Jersey cattle and the beef breeds of Charolais, Limousin, Hereford and Aberdeen Angus, largely reside in the cooler and temperate regions of the world such as Europe, parts of Australia, New Zealand, northeastern Asia and parts of Africa. Humped cattle favour the hotter regions of the world where *Bos taurus* would struggle to cope. With a fatty hump on their shoulders, drooping ears and a large dewlap, the versatile *Bos indicus* can be seen pulling farm equipment as well as providing meat, milk, hides and their dung for fuel and manure. The multi-functionality of domestic cattle sealed their fate as valuable commodities for intensive animal agriculture, with long-term welfare implications.

Before they became walking larders, cattle were the adored representatives of the gods in some cultures (Clutton-Brock, 2012). For example, Hathor, an ancient and important cow goddess in early Egyptian civilisations, carried a sun disc between her horns. However, being worshipped did not protect against early death, with some ancient tomb paintings depicting sacrificial bulls being handed over to the gods for slaughter. The ancient Israelites perceived cattle as sacrifices, workers or providers of their body parts, none of which ended well for the animals.

A lingering belief in the sanctity of domesticated cattle remains in some cultures. Among those who identify as Hindu, some perceive them as worthy of reverence and therefore to be cared for and protected (Valpey, 2020). However, India, where 80 per cent of the population are Hindu but only 20 per cent are vegetarian, ranked fifth among nations with the greatest number of cattle slaughtered, suggesting that the legacy of India's special regard for the sacred meaning of these animals may be changing (Challenger, 2021). Meanwhile, on Norfolk Island, a tiny island in the South Pacific Ocean, cows provide an interesting tourist feature as they enjoy the permission granted to them in the 1800s to wander freely with right of way over traffic.

Beef consumption worldwide is the third highest among farmed animals, behind poultry and pork (Siddiqui et al., 2023). The global live cattle population was estimated at over 1.57 billion in 2023 (Shahbandeh, 2025), with the United Nations Food and Agriculture Organisation (2023) estimating that over 300 million are slaughtered each year to meet human demand for meat, leather and other by-products. This equates to almost 900,000 every day.

CATTLE IN CONTEMPORARY SOCIETY

As with many other animals, domestication brought about changes in different parts of the cow's body and brain. There was a reduction in overall body size compared to the massive aurochs, a reduction in horn size and shape and the selection of personality traits to make them more accepting of intensive farming and milking (Clutton-Brock, 2012). Castration of bulls for human safety sought to turn an aggressive bull into a submissive work animal, but this also resulted in increases in body fat and changes to bone dimensions. Brain size diminished, possibly reflecting changed living environments with less need to be alert and forage and no mental stimulation. It is widely accepted that domesticated animals never show an increase in brain size compared to their wild relatives as domestication converted intelligent, social animals into bored commodities for human use (Ballarin et al., 2016).

Observe a group of cattle in a field or other natural setting, and their behaviour suggests contentment with life as they graze, ruminate, wander about and avoid confrontation. Meanwhile, calves kick up their heels and run for the sheer enjoyment of being alive. Believing that these sentient animals lack capacity to think beyond instinctual behaviours provides a strategy to appease the guilt and tension of cognitive dissonance, but in reality, it is domestication that has contributed to cattle who are not allowed to be smart nor to develop their natural emotional and behavioural capacities.

Many also do not live long enough to attain the wisdom and experience normally associated with age. Bobby calves are slaughtered soon after birth, veal calves at six to eight months, beef cattle before 18 months and dairy cows around four to seven years when no longer producing sufficient milk to be economical—a vast reduction in their potential natural lifespan of 15–25 years (Farm Transparency Project, 2024). Cattle can learn to do most of the things that people teach their beloved companion dogs: to stay, spin, fetch and play with balls, as well as additional farm specific tasks such as pressing buttons to gain entry to buildings.

Even without training, cows have used observational learning to work a pump handle to bring up water from a well and press a lever for extra food (Comstock, 2020). Instead of being given the opportunity to exercise this natural intelligence, cows confined to intensive farm settings must contend with lack of mental stimulation, suffering and eventual slaughter.

A major trauma in the lives of many dairy cows is separation from their newborn calves. Early separation remains the norm in some countries, supported by rationales such as reducing the stress of later separation after stronger bonding, protecting the mother from teat damage and increasing the saleable milk from the mother (Neave et al., 2022). While research increasingly confirms the common-sense observation that when forcibly parted from their newborn calves, dairy cows demonstrate negative emotional responses such as an increase in visible eye whites (Proctor & Carder, 2015), there are mixed results as to the least distressing age to separate mother and calf.

Despite calves having capacity to begin bonding with their mothers as early as five minutes after post-natal contact, separating cow and calf within 24 hours is still perceived as beneficial in reducing bonding and the resultant separation distress (Weary & Chua, 2000). Researchers Daniel Weary and Beverly Chua (2000) reported that while cows and calves expressed vocal and other behavioural responses to separation at six hours, one day and four days postpartum, there was greater distress at later separation. Mothers vocalised more at four-day postpartum separation and their calves were more restless. They concluded that the time of separation needed to be balanced with other factors such as the benefits of calves receiving their mothers' early colostrum-rich milk, which could not be sold anyway. While these studies debate the best time to separate mother and calf, they failed to address the fundamental ethical implications of a process that disadvantages a subordinate species so that the dominant species can consume the milk and milk products that are not rightfully theirs.

By nature, cows choose to live in herds, and the mother seeks isolation to give birth before returning to the group. Mothers nurse their young five to nine times daily for the first few weeks of life, explaining why producers must separate mother and calf if there is to be any economically viable milk left for human consumption. Mothers care for their calves until 6–14 months of age, but the strong mother-child bond can last for years and includes social learning and affiliative behaviours, which are prevented by early separation on dairy farms. Despite some countries promoting extended cow-calf time, partly in response to growing consumer awareness of animal welfare (Vaarst et al. 2020), the ultimate outcome remains unchanged: bull calves face premature death, either soon after birth or several months later for veal, or heifer calves are reared to replace adult milk-producing cows until they too are spent like their mothers.

Duncan, the Jersey calf, was destined for early slaughter, and his carcass segmented into veal meat cuts prized for their mild flavour and tenderness. However, his life changed dramatically when rescued by Goodheart Farm Animal Sanctuary on the border between Worcestershire and Shropshire in the United Kingdom.

Duncan and the Jersey boys—Goodheart Farm Animal Sanctuary, United Kingdom

Duncan never knew his mother. They were separated shortly after his birth before he had chance to start building a healthy immune system. Duncan, along with five others, was destined to end up on a veal rearing farm. It was the start of a long cold winter, and three-week-old Duncan had grown so fragile that he could no longer stand.

These days it seems unbelievable that the noisiest, cheekiest cow at Goodheart Farm Animal Sanctuary was once so weak and vulnerable that he had to be carried from the trailer to his new home. Of the six male Jersey calves rescued that cold, wet November in 2017, Duncan's condition was the most concerning and his very survival in doubt. Cocooned in two little calf jackets to keep his frail body warm, Duncan's fight for life began in the sanctuary utility room, where he was spoon fed medicine by the team who were determined to give the young calf a chance to reach adulthood.

And it worked. The tiniest calf with the huge dark eyes beat the odds and together with his five friends grew strong enough to leave the warm barn six months later and meet the world. Instead of heading for the slaughter yard, the six strong, happy and healthy "Jersey boys"—Duncan, Jerry, Woody, PJ, Jacob and Geoffrey—were free to race around a field for the first time in their lives. As a group, they took off, jumping with joy as they explored their new home, grazed in the sunshine or stopped to greet a friendly pig. The Jersey boys were home and loving every minute of life and freedom rather than their previous destiny of being slaughtered and served up on a dinner plate at that age.

Duncan's unique personality did not take long to emerge. With his huge expressive brown eyes and a tussle of curls on top of his head, the boy was irresistible to the team and sanctuary visitors alike. Duncan still carried the scars of his rough start to life with a head tilt, possibly the legacy of an old ear infection and an underbite that made him instantly recognisable, but nothing could hold him back. Described by the Goodheart team as "a gentle soul who loves to interact with his human friends", Duncan's loud vocalisations guaranteed attention. The first and loudest to greet visitors to his paddock, Duncan remained constantly on the look-out for a good head scratch, or even better, a good ear scratch combined with a lot of loving fuss.

In the 12 months December 2023 to December 2024, between 1,000 and 2,000 calves less than one year old were slaughtered every week in the United Kingdom (Department for Environment, Food and Rural Affairs, 2025). These calves contributed to approximately 85 grams of veal and beef purchased per person per week to consume at home (Bedford, 2024). Duncan is now fully grown and will never be part of these statistics. Despite the fear and trauma in his early life and resultant negative impact on his immune system, Duncan survived. The psychological effects may remain embedded in his long-term memory, but within the safety of the sanctuary, there are no traumatic visual, auditory or olfactory triggers reminiscent of Duncan's early start to life in the industrial farm system. He and his friends are free to express their unique personalities and live their best lives at Goodheart Farm Animal Sanctuary.

Figure 3.1 Duncan, with his expressive brown eyes and a tussle of curls on top of his head. (Image supplied by Goodheart Farm Animal Sanctuary and reproduced with their permission.)

Beef cattle

Beef cattle also experience a range of intensive farming systems depending on their location, with varying negative welfare impacts (Salvin et al., 2023). From around the 1920s, intensive feeding of beef cattle in large-scale commercial feedlots spread around the world from the USA and Canada to Australia, New Zealand, South America, South Africa, parts of Asia and the European Union. A South African feedlot capable of holding 150,000 beef cattle is cited as the largest in the Southern Hemisphere, while USA, Mexico and Brazil have the overall largest number of confined cattle (Vale et al., 2019). Many beef cattle spend at least a portion of their lives in the confinement of feedlots when they are moved from grazing pastures to be fattened up for slaughter. Feedlots are an area of fenced or walled open ground, either indoors or outdoors, containing feed and water troughs. The floors are sloped to allow drainage, with some sort of bedding included to ameliorate muddy conditions. Indoor confinement areas may have solid floors or slatted floors to allow manure to drop through.

Feedlots vary, but most have fundamental welfare issues when beef cattle as young as five to eight months are confined in close quarters. Apart from the obvious ethical issues associated with confining animals to deliberately fatten them up for slaughter, cattle can experience increased stress levels from high density confinement, rapid spread of diseases such as bovine respiratory disease and bovine foot rot from excessive standing on mud (Vale et al., 2019). In warmer climates, heat stress can be compounded by the high-energy diets that generate metabolic heat during the fattening process (Salvin et al., 2023). While restricting feed can reduce an animal's body temperature during extreme heat events, the flow-on effect of slower daily weight gain and longer times to reach marketability impact profitability and introduce additional welfare implications.

Brazil has attracted considerable attention for having the largest cattle population in the world (over 230 million head), as well as being one of the largest beef exporters in the world (Ferreira et al., 2024). Brazil also has the third highest number of confined cattle in the world (Vale et al., 2019), typically restricted to the last three months of their life. Climate projections report increasing temperatures in some regions, with similar projections in parts of the USA, the third largest cattle population in the world. As global warming extends to more and more locations, both grazing and confined cattle will experience welfare issues associated with heat stress. In the Northern Great Plains (NGP) region of the USA, temperatures exceeding 55 degrees Celsius (131 degrees Fahrenheit) in unshaded feedlots on hot sunny days have been recorded (Derner et al., 2018).

Whether overheated or not, there is a risk of antagonistic behaviour among high-density feedlot cattle. Environmental enrichment is increasingly recognised as necessary for confined animals such as rodents in research laboratory settings and intensively housed chickens, but minimal studies have been conducted among feedlot cattle. Inclusion of a scratching brush within the pen to reduce stereotypies and aggressive behaviours has been explored, with some evidence of increased aggressive behaviours when densely packed cattle compete to reach it (Salvin et al., 2023). For an animal capable of complex cognitive manipulations, attachments and emotional experiences, confinement in a feedlot or dairy farm environment can be distressing.

FEELINGS AND EMOTIONS

Cattle have relatively large brains—for example, around 480 grams (one pound) in a 600-kilogram (1,322 pound) animal—with a highly folded cerebral cortex, but relatively low neuronal density (Ballarin et al., 2016). As with other mammals, including humans, folding allows a larger cortical surface area to fit within the skull and correlates with greater cognitive functionality fitted into a smaller space. However, there are some gaps in understanding of the cellular composition of their central nervous system. Both cattle and humans have a similar gestation period of around nine months, but the unborn calf's central nervous system matures comparatively early during pregnancy to allow the newborn to quickly stand, move and relate to the external world (Peruffo & Cozzi, 2014). Early maturation and perception suggest that newborn bobby calves may be acutely aware of the stressors associated with separation from their mothers and subsequent slaughter.

Cattle have a well-developed sense of smell, as evidenced by their large olfactory bulbs, hippocampus and related structures, and their motor pathways for quadrupedal locomotion show extensive development of the extrapyramidal multi-synaptic tracts (Cozzi et al., 2020). This has allowed the bovine brain to be used as a research model for conditions such as transmissible spongiform encephalopathies in humans and other animals. However, a well-developed sense of smell also suggests that the odours of blood and excrement associated with the traumatic truck roll-over in suburban Melbourne had the potential to become embedded in the steers' memories via the hippocampus, which plays an important role in both memory formation and emotional processing.

Fear can affect hormones with resultant physical effects. Dairy cows experiencing different types of handling (that is, slaps and shouts from one handler and more gentle handling from another) retained more milk when in the presence of the rough handler (Fraser & Weary, 2022). The researchers hypothesised that the neuro-endocrine responses triggered by rough handling generated fear, which in turn impacted the actions of the milk-releasing hormone oxytocin. While fear and rough handling are obviously detrimental to the cow, the economic implications of reduced milk yield should provide sufficient motivation to reduce such stressors, and yet this is not always the case.

Body function in cattle is altered in response to the stressor via the hypothalamic-pituitary-adrenal (HPA—hypothalamus, the anterior pituitary gland and the adrenal glands) axis (Earley et al., 2023), a response common across a range of farmed animals in intensive animal agriculture. The HPA axis response to stress commences within the animal's brain, eventually resulting in the secretion of adrenocorticotropic hormone (ACTH) from the anterior pituitary. This acts on the adrenal gland, which regulates glucocorticoid in their blood. While glucocorticoid helps short-term survival of cattle, prolonged exposure can result in metabolic, immune and psychological dysfunction. Cortisol is the main active ingredient of glucocorticoid and has been used as a biomarker of stress in calves and cattle when weaning, transporting, castration, grouping and regrouping. Inevitably, unknown confounding variables make it hard to identify one stressor, especially when factoring in handling for the blood test, time of day and temperature, but mounting evidence suggests that these sentient animals with complex brain structures do experience stress, and much of it can be related to maximising production output in industrial farm systems.

As with other mammals (and humans), the two hemispheres of cattle brains are specialised to process information from the environment in different ways. The right hemisphere processes inputs from new or challenging stimuli, and the left hemisphere processes inputs from predictable stimuli. With the usual cross-over of the optic nerve, information from the right eye is processed in the left hemisphere and from the left eye in the right hemisphere (Haskell & Langford, 2023). Cumulative evidence suggests that subordinate animals viewed other cattle and unfamiliar humans with their left eye to be processed in their right hemisphere as a new or challenging stimuli. Familiar humans were viewed through the right eye to be processed in the left hemisphere as predictable stimuli. This emotional-driven laterality in the eyes provided a non-threatening means to measure emotional responses as to how threatening one animal finds other cattle, new objects or human handlers. These studies, combined with knowledge of cattle neuroanatomy, support the notion that cattle assess and respond emotionally to the potential outcomes of a situation—even what the future may hold—thus raising questions about the moral right of humans to deprive them of that future (Comstock, 2020).

In human psychology, it is widely accepted that relentless pain, fear or other unpleasant states not only impact mood, but set up a downward spiral where all information is interpreted pessimistically. This cognitive bias perpetuates and deepens the low mood until chronic depression sets in. Cognitive bias has also been demonstrated in a range of animals, including dairy calves. Drawing on researchers Emma Harding, Elizabeth Paul, and Michael Mendl's (2004) early studies where rats were trained to respond to different sounds that led to either a food reward or an unpleasant noise, dairy calves were taught to discriminate between different colours on a monitor associated with a positive or negative cue (Fraser & Weary, 2022). After undergoing the painful process of hot-iron disbudding to prevent horn buds from developing, they showed a pessimistic response to an intermediate colour cue. When these same calves were separated from their mothers a few weeks later, this pessimistic response remained evident, suggesting that long-term mood states can be set in motion by routine farm practices that fail to respect the feelings of these sentient animals.

Cattle develop social structures within their communities, choosing to spend time with specific individuals with whom they form bonds and special relationships. Their excellent memories and ability to identify up to 70 other herd members increase the likelihood of negative emotional and physical consequences when separated from their social groups of choice. Association with preferred familiar companions can have a calming effect on dairy cows, in contrast to the evidence of increased heart rate, vocalisations and plasma cortisol concentrations among heifers when socially isolated (Boissy & Le Neindre, 1997). These negative reactions subsided when reunited with their peers. While all these findings are indicative of distinct traits and emotions, no different to a companion animal dog or cat, very few urban dwellers will ever get the chance to see a cow chasing a ball or running and jumping just for the sheer joy of being alive.

In 2020, researchers Heather Neave and colleagues sought to describe the development and stability of personality traits from birth to adulthood in Holstein dairy cattle. Interest in the personality of domesticated farmed animals has largely centred around potential links to individual productivity, amenability to confinement, health and welfare, with commonly researched traits including the likelihood of them taking risks when exposed to something new (boldness); their behaviour when seeking information about the environment (exploration) and their propensity to seek out and remain close to conspecifics (sociability). All these traits can be inferred by observing behavioural responses, which do little to truly understand their personality beyond what is helpful for farming. Dairy cattle undergo many important developmental and psychological changes that have been accelerated in commercial farm settings. There is immediate separation from the mother, early weaning to solid food at two months, puberty before 12 months and pregnancy and lactation by 24 months. Neave and colleague's study looked at stability of personality traits across four periods of development, finding that cows had relatively stable personality traits during early and late development periods (that is, in the pre- to post-weaning periods and from puberty to lactation). Individual consistency of behaviour and personality traits was poor across the period of puberty.

The importance of these outcomes was linked back to boosting productivity among dairy cows rather than recognition and acceptance of these farmed animals as individuals with unique personalities. Where personality was acknowledged, it was within the context of feeding behaviours and performance on-farm. Reactive cows to novelty or stressful situations recorded impaired feeding behaviour, feed intake and immune function. If reactive when restrained in a crush, there were greater financial implications with reduced growth, carcass quality, reproductive performance and reduced milk production. Selectively breeding out these personality traits could therefore be beneficial for production, but further commodify cows to fit human needs.

Negative emotions such as fear, apprehension, stress and anxiety have also been linked to poor financial outcomes when reducing appetite and the resultant reproductive, production and immune consequences. This risk was sufficient motivation to develop methods to determine if cattle were suffering from prolonged and chronic hyperarousal in order to intervene for productivity reasons. Tests included flight speed, crush score, forced lateralisation test, human approach test, cortisol concentration (usually in blood plasma, milk, saliva, urine or faeces), eye white and ear posture (Uddin et al., 2023).

Mental Experiences, Domain 5 of Mellor's Five Domains, provides an assessment of an animal's overall affective experience, which can be shaped by the animal's fear-induced subjective perceptions and interpretations of their world. Where human interactions during handling, transport and slaughter are associated with fear, stress and anxiety, the cow's lack of agency can reduce positive emotional states. Inevitably, both the positive and negative emotional lives of farmed animals are affected by their interactions with the humans who control every aspect of their lives. Recognition that animal welfare must include positive experiences such as physical and emotional comfort and pleasure can create tension between animal welfare science and economics, especially when the four defining features of a "good life" become costly in oversized intensive settings. These features, defined by researchers Elizabeth Rowe and Siobhan Mullan (2022) as positive emotions, positive affective engagement, good quality of life and happiness, can be found in abundance at farmed animal sanctuaries, but are lacking in cattle confined to a dairy farm or a feedlot.

Positive emotions arise when animals have the opportunity to perform behaviours that they are motivated to perform—and enjoy—such as grazing, maternal care and play. These experiences result in positive affective engagement, and when coupled with resources to achieve positive physical experiences, including physical and thermal comfort, the ability to rest, eat and drink, are assumed to result in a good life (Rowe & Mullan, 2022). Achieving this good life, where positive welfare substantially outweighs negative welfare and where all the animal's needs and most of their wants are met, comprises a standard that few intensive farm settings can achieve and few laws and codes can enforce. A good life recognises choice to determine what the animal wants as opposed to what is imposed on them to meet human needs. Choice is never one-size-fits-all for sentient beings with personalities and preferences. There are choices driven at a species level, but there are also individual variations in preference, something that is difficult, if not impossible, to achieve in intensive animal agriculture. With no current gold standard to consistently assess and validate a cow's emotional state, animal welfare legislation will always fall short of protecting the individual's right to choice at any level. In contrast, farmed animal sanctuaries strive to create environments where some choice is possible, emotional capacity is respected and cows and other residents can live their best lives.

Snoopy, living the good life—Austin Farm Sanctuary, USA

A long-held narrative is the assumption that only the fittest animals survive—natural selection weeds out those vulnerable, weak or disabled animals who then go off to die or be killed. This assumption, combined with the economic implications in supporting a disabled farmed animal, has resulted in another narrative emerging—the "better off dead" narrative that has made disability and death synonymous in intensive farm systems. Ableism is found among companion animals as well, where the so-called "less adoptable pets" with physical or psychological disabilities can be overlooked or euthanised. Therefore, it should come as no surprise that when a six-month-old blind calf found himself at a cattle auction, his future looked grim. Born on a cattle farm to be raised for beef, the farmer was unable to care for the blind calf's perceived special needs and sent him to be sold off for meat at auction. For young calves separated from their mothers, cattle auctions are loud, busy, smelly and terrifying places. For a young blind calf, the terror must have been unimaginable as he roamed sightlessly through the auction yard.

The young calf's tangible terror caught the attention of a teenage boy who chose to purchase the helpless infant and create a new beginning for him. Children from as young as one or two years have natural empathy for all animals, but as they develop into young adults, a dichotomy can develop between attachment to companion animals and less regard for "food" animals (Piazza, 2020). Psychologist Melanie Joy (2009) described a collective dissociation of meat from animals, which prevented people from seeing the suffering of farmed animals and acting upon universal impulses for compassion. She argued that more people needed to "bear witness" to the suffering of farmed animals so social practices were more aligned with social values. Bearing witness to the distress of one young calf at the auction yard was sufficient motivation for the 15-year-old teenager to save a life.

Purchasing Snoopy was only the first step. The teenager and his mother then had to find somewhere willing to take on this "disabled" farmed animal for the rest of his life—somewhere like Austin Farm Sanctuary in Texas, USA.

In 2016, two goats and a pig became the catalyst for a sanctuary that would eventually provide safety and security to over 250 rescued animals on a daily basis. Austin Farm Sanctuary co-founders, Chris and Angela, describe their mission very simply: "To rescue and provide sanctuary to farmed animals and inspire change in how society views and treats these wonderful souls" (Austin Farm Sanctuary, 2023). Austin Farm Sanctuary started in South Austin, but like many sanctuaries, once the animals kept arriving, they soon outgrew their modest beginnings. There was a second move as the sick and abandoned animals multiplied, then a third and final move in 2022, which included greater room for more animals and the building of a Medical Barn.

In 2020, the arrival of four-week-old Buddy, a goat born blind and left struggling to locate his mother and keep up with the herd at the goat farm, highlighted the risks that disabled farmed animals faced in commercial enterprises. Buddy's arrival eventually led to the establishment of designated spaces like "Buddy's Buds" and "Copper's Crew" in the new location. Copper's Crew, the medical barn paddock for sickly or disabled cattle, provided the perfect new start in life for young Snoopy.

Most animals can adjust to lack of vision if given the chance, and Snoopy was no exception. The first step was to work out what if anything could be done about Snoopy's eyes, knowing that whatever the outcome, he would receive a lifetime of care and love at Austin Farm Sanctuary. The aim was to ensure that the young calf would never again experience the fear, pain and uncertainty of his early start to life, instead replacing these negative stressors with friendships, healthiness, security—and a regular dose of snuggles, head scratches, sunset cud chewing and bananas.

And that is exactly what happened. Snoopy joined the small herd of disabled cows, and for the next six months, Chris, Angela and the medical team worked to further understand his eyes, provide him with pain relief and make a plan to heal him in whatever way was needed or possible. By 2023, the pressure in Snoopy's eyes was increasing, causing pain and risking further complications, and so he was booked in for a medical evaluation and possible surgery to remove one or both of his eyes. There were celebrations all round when the ophthalmology consultation prior to surgery showed that the pressure had decreased in his eyes and he was not in pain. His vision would never return, but with eyedrops and

Figure 3.2 Snoopy, always ready for head scratches and snuggles. (Image supplied by photographer Sarah Throop, Austin Farm Sanctuary and reproduced with permission.)

the support of human and animal friends alike, he could continue living the cow dream life at Austin Farm Sanctuary. Before long, three more blind calves joined Snoopy in Copper's Crew: Chance, who was found stuck in mud up to his neck, and Linus and Boo, who were born at cattle ranches and losing their sight after getting sick and developing sepsis.

Snoopy is testament to the resilience and forgiving nature of farmed animals who may meet the human definition of disabled but refuse to be defined by a socially constructed label. Many of the animal residents at Austin Farm Sanctuary had a rough start to life, but their narratives were rewritten when given the opportunity to express their unique personalities and preferences in safety and look forward to a future that once seemed impossible. As Snoopy is caught snoring gently with his head on Chris' lap, it is clear that one boy's act of kindness has changed this sentient animal's life forever.

ETHICAL IMPLICATIONS

While it is important to understand the functioning of cattle brains in order to assist with animal welfare, this is only the first step to understanding their emotional capacity. Learning about the brain and how cattle perceive the world and the emotions they feel can become uncomfortable when a disturbing moral dilemma emerges. Cattle are commodified in industrial farm systems to meet human demand, whether that be milk, cheese and yoghurt or the massive number of hamburgers available from fast food outlets worldwide. The cognitive dissonance of recognising emotional capacity in cattle while simultaneously craving the products from their bodies can be alleviated when accepting at face value the information a producer like McDonald's Fast-food Corporation presents to consumers on their website:

> The Company requires that, for raw material suppliers of beef, chicken and pork, slaughterhouses follow a practice of humanely stunning animals prior to slaughter (except to the extent prohibited for religious reasons). We believe stunning animals ensures insensibility before slaughter.
>
> *(McDonald's, 2025)*

Describing beef and other animals as "raw materials" that are "humanely" stunned prior to slaughter distances the reader from any emotive connection between sentient animals and death and perpetuates the consumption of guilt-free hamburgers. Language is a powerful tool to allay cognitive dissonance, and marketing and advertising machines make good use of this method to distance the reality of slaughter from the idyllic life and death they describe for their raw materials.

Farmed animals are more than raw materials. Animal advocates, ethologists, neuroscientists and philosophers have worked hard to show that animals have feelings. The 2012 *Cambridge Declaration on Consciousness* is testament to this, and yet farmed animals continue to be cruelly used and abused to fulfil human needs. Marc Bekoff (2024), a famous ethologist and animal behaviourist, commented that there remains a separation of "us" (humans) from "them" (other animals), but stressed the importance of paying attention to successes on behalf of nonhuman animals, not just the challenges and harms still being done. The story of Bruiser and Oreo from Forever Friends Animal Rescue in Australia is one of those successes despite a series of challenges along the way.

Bruiser, overcoming challenges—Forever Friends Animal Rescue, Australia

Animals caught up in intensive animal agriculture are often forced to suffer in silence because no one has the time—or inclination—to listen to them. They remain voiceless and powerless, until someone cares enough to look into their eyes and speak for them. At just two weeks of age, calves Bruiser and Oreo found themselves voiceless and alone after experiencing separation from their mother, transported to a loud, smelly and terrifying abattoir and confusion as to what would happen next in their short lives.

Their future changed dramatically when they were rescued from death by a caring person and eventually arrived at Forever Friends Animal Rescue in Victoria, Australia. Located on 40.5 hectares (100 acres) in the beautiful Yarra Valley, Forever Friends was founded in 2011 with a mission to save and transform the lives of homeless, neglected and abused animals. Farmed animals, ex-racehorses, dogs and cats—all have had their lives transformed at the sanctuary, or with one of the dedicated foster carers and volunteers who are just as determined as founder Saskia Adams to make a difference in the lives of these sentient beings.

Bruiser and Oreo thrived at Forever Friends as they matured into two gentle giants with a passion for people and the treats they found hidden in their visitors' pockets or bags. Always polite, the boys traded kisses with their huge rough "cat" tongues for the treats they discovered on the humans who grew to love them. When the diminutive 18-month-old Friesian cow Sally was rescued from a dairy farm, Bruiser and Oreo welcomed her to the sanctuary, and the three unique individuals formed a tight-knit group whose futures were secure at Forever Friends.

At nine years old, Bruiser developed a large rapidly growing tumour on his right eye. Removal of the eye promised the best outcome for Bruiser, and after a successful surgery, he was back to being the friendly comical giant always on the look-out for pats and snacks or hanging out with his friends.

However, Bruiser's challenges were not yet over. A series of small cancerous spots on his left eye were kept at bay by routine freezing until one appeared in an inaccessible location. Bruiser's left eye was in jeopardy, and decisions had to be made. Snoopy, whose story appeared earlier in this chapter, had never known sight and adapted to a sightless existence as the norm. But Bruiser had lived for almost a decade with vision, and the life change would be significant. Bioethicist Jessica Pierce (2019) drew attention to the need to respect consent, assent and dissent of an animal when making decisions and to avoid a paternalistic approach that failed to accommodate their rights and abilities. Bruiser was unable to articulate his consent or dissent for a procedure that would save his life but take his vision, and so the Forever Friends team looked for other ways to understand the thoughts and feelings of their beloved friend.

There seemed no doubt that Bruiser remained interested in the world around him and retained a love for life. His quiet dignity and trust as he accepted 12 months of treatments to his left eye and still came back for snacks and kisses suggested that Bruiser was not yet ready to give up on life. He was eating, drinking and socialising, and the removal of his eye would allow him to continue this for many more years to come. At ten years, Bruiser was not elderly, but he had lived far longer than the two-week old calf's original life trajectory. Proceeding with surgery would prevent long-term physical suffering and eventual euthanasia as the cancer advanced—but he would then meet the human definition of disabled and the stereotypes associated with that construct. As with Snoopy, Bruiser risked falling victim to the "better off dead" narrative that has made disability and death synonymous among farmed animals.

When contemplating animal euthanasia in his pivotal book *The Case for Human Rights* (2004), animal ethicist Tom Regan succinctly debunked what he called paternalistic euthanasia: "To kill an animal 'for his own good' is paternalistic: we impose *our* will and *our* judgment on the animal, for the animal's own good, as we conceive it" (p. 115). Regan added that part of the flawed rationale for paternalistic euthanasia of animals needing expensive treatments in shelters was to make better use of limited resources for other animals. Bruiser's surgery would be expensive for Forever Friends, and arguably, there were many other ways to use these funds than on a ten-year-old cow whose prognosis was blindness. However, even without words, Bruiser's will to live was undeniable, and the decision was made: Bruiser would have the surgery, and his brother Oreo would be enlisted to help out.

In preparation for Oreo's new role, he started wearing a bell around his neck to accustom Bruiser to the sound. This would ensure Bruiser stayed out of trouble—and importantly, facing in the right direction for snacks—once the surgery was over. The plan was successful, and just hours after surgery, Bruiser could be seen following Oreo around, but importantly, also finding things himself. Watching Bruiser contentedly chewing his cud, Saskia and the team at Forever Friends marvel at the adaptability of this gentle giant who refuses to allow blindness—and disability—define his life.

Figure 3.3 Bruiser (left) and Oreo waiting eagerly for visitors and snacks. (Image supplied by Forever Friends Animal Rescue and reproduced with their permission.)

REFERENCES

Austin Farm Sanctuary (2023). *Our mission*. https://www.austinfarmsanctuary.org/our-mission

Ballarin, C., Povinelli, M., Granato, A., Panin, M., Corain, L., Peruffo, A., & Cozzi, B. (2016). The brain of the domestic *Bos taurus*: Weight, encephalization and cerebellar quotients, and comparison with other domestic and wild Cetartiodactyla. *PLoS One, 11*(4), e0154580. https://doi.org/10.1371/journal.pone.0154580

Bedford, E. (2024). Weekly UK household consumption of beef and veal 2006–2023. *Statista*. https://www.statista.com/statistics/284343/weekly-household-consumption-of-beef-and-veal-in-the-united-kingdom-uk/#:~:text=In%202022%2F23%2C%20an%20average%20of%2085%20grams%20of,for%20consumption%20at%20home%20in%20the%20United%20Kingdom

Bekoff, M. (2024). *The emotional lives of animals: A leading scientist explores animal joy, sorrow, and empathy—and why they matter*. New World Library.

Boissy, A. & Le Neindre, P. (1997). Behavioural, cardiac and cortisol responses to brief peer separation and reunion in cattle. *Physiology & Behavior, 61*(5), 693–699. https://doi.org/10.1016/S0031-9384(96)00521-5

Challenger, M. (2021). *How to be animal. A new history of what it means to be human*. Canongate Books Ltd.

Clutton-Brock, J. (2012). *Animals as domesticates: A world view through history*. Michigan State University Press.

Comstock, G. (2020). Bovine prospection, the Mesocorticolimbic pathways, and neuroethics: Is a Cow's future like ours? In L.S.M. Johnson, A. Fenton, & A. Shriver (Eds.), *Neuroethics and nonhuman animals* (pp. 73–98). Springer.

Cozzi, B., Bonfanti, L., Canali, E., & Minero, M. (2020). Brain waste: The neglect of animal Brains. *Frontiers in Neuroanatomy, 14*, 573934. https://doi.org/10.3389/fnana.2020.573934

Department for Environment, Food and Rural Affairs (2025). *Latest cattle, sheep and pig slaughter statistics*. https://www.gov.uk/government/statistics/cattle-sheep-and-pig-slaughter#home-killed-meat-production

Derner, J., Briske, D., Reeves, M., Brown-Brandl, T., Meehan, M., Blumenthal, D., Travis, W., Augustine, D., Wilmer, H., Scasta, D., & Hendrickson, J. (2018). Vulnerability of grazing and confined livestock in the Northern Great Plains to projected mid- and late-twenty-first century climate. *Climatic Change, 146*, 19–32. https://doi.org/10.1007/s10584-017-2029-6

Earley, B., Buckham-Sporer, K., O'Loughlin, A., & Johnston, D. (2023). Physiological and immunological tools and techniques for the assessment of cattle welfare. In M. Haskell (Ed.), *Cattle welfare in dairy and beef systems. A new approach to global issues* (pp. 165–202). Springer. https://doi.org/10.1007/978-3-031-21020-4

Farm Sanctuary (2021). *40 cows flee L.A. slaughterhouse; compassion brings two to sanctuary*. 40 Cows Flee L.A. https://www.farmsanctuary.org/news-stories/40-cows-escape-slaughterhouse/

Farm Transparency Project (2022). *Painful procedures.* https://www.farmtransparency.org/kb/food/cattle/painful-procedures

Farm Transparency Project (2024). *Age of animals slaughtered.* https://www.farmtransparency.org/kb/food/abattoirs/age-animals-slaughtered

Farm Transparency Project (2025). *End dairy slaughter.* https://www.farmtransparency.org/campaigns/end-dairy-slaughter

Ferreira, N.C.R., Andrade, R.R., Ferreira, L.N., & Grandin, T. (2024). Climate change impacts on livestock in Brazil. *International Journal of Biometeorology, 68*, 2693–2704. https://doi.org/10.1007/s00484-024-02778-3

Forever Friends Animal Rescue (n.d.). *About us.* https://www.foreverfriends.org.au/about/

Fraser, D. & Weary, D.M. (2022). Applied animal behavior and animal welfare. In J.J. Bolhuis, L.A. Giraldeau, & J.A. Hogan, *The behavior of animals, 2nd edition: Mechanisms, function, and evolution* (pp. 251–280). John Wiley & Sons, Inc.

Grandin, T. (2001). Welfare of cattle during slaughter and the prevention of nonambulatory (downer) cattle. *Journal of the American Veterinary Medical Association, 219*(10), 1377–1382.

Haskell, M.J. & Langford, F.M. (2023). Using behaviour to understand and assess welfare in cattle. In M. Haskell (Ed.), *Cattle welfare in dairy and beef systems. A new approach to global issues* (pp. 15–54). Springer. https://doi.org/10.1007/978-3-031-21020-4

Joy, M. (2009). *Why we love dogs, eat pigs, and wear cows.* Conari Press.

Lidfors, L. (2022). Parental behavior in bovines. In G. González-Mariscal (Ed.), *Patterns of parental behavior. Advances in neurobiology* (177–212). Springer. https://doi.org/10.1007/978-3-030-97762-7_6

McDonald's Fast-food Corporation (2025). *Animal health & welfare.* https://corporate.mcdonalds.com/corpmcd/our-purpose-and-impact/food-quality-and-sourcing/animal-health-and-welfare.html

Neave, H.W., Costa, J.H.C., Weary, D.M., & von Keyserlingk, M.A.G. (2020). Long-term consistency of personality traits of cattle. *Royal Society Open Science, 7*, 191849. https://doi.org/10.1098/rsos.191849

Neave, H.W., Sumner, C.L., Henwood, R.J.T., Zobel, G., Saunders, K., Thoday, H., Watson, T., & Webster, J.R. (2022). Dairy farmers' perspectives on providing cow-calf contact in the pasture-based systems of New Zealand. *Journal of Dairy Science, 105*, 453–467. https://doi.org/10.3168/jds.2021-21047

Peruffo, A. & Cozzi, B. (2014). Bovine brain: An in vitro translational model in developmental neuroscience and neurodegenerative research. *Frontiers in Pediatrics, 10*(2), 74. https://doi.org/10.3389/fped.2014.00074. PMID: 25072040; PMCID: PMC4090595.

Piazza, J. (2020). Accomplishing the most good for animals. In K. Dhont & G. Hodson (Eds.), *Why we love and exploit animals* (pp. 121–136). Routledge.

Pierce, J. (2019). The animal as patient. Ethology and end-of life care. *Veterinary Clinics of North America: Small Animal Practice, 49*, 417–429. https://doi.org/10.1016/j.cvsm.2019.01.009

Proctor, H.S. & Carder, G. (2015). Measuring positive emotions in cows: Do visible eye whites tell us anything? *Physiology & Behavior, 147*, 1–6. https://doi.org/10.1016/j.physbeh.2015.04.011
Regan, T. (2004). *The case for animal rights.* University of California Press.
Rowe, E. & Mullan, S. (2022). Advancing a "good life" for farm animals: Development of resource tier frameworks for on-farm assessment of positive welfare for beef cattle, broiler chicken and pigs. *Animals, 12*, 565. https://doi.org/10.3390/ani12050565
Salvin, H., Schwartzkopf-Genswein, K., Lee, C., & Colditz, I. (2023). Welfare of beef cattle in intensive systems. In M. Haskell (Ed.), *Cattle welfare in dairy and beef systems. A new approach to global issues* (pp.165–202). https://doi.org/10.1007/978-3-031-21020-4
Shahbandeh, M. (2025). Cattle population worldwide 2012–2023. *Statista.* https://www.statista.com/statistics/263979/global-cattle-population-since-1990/
Siddiqui, S.A., Gerini, F., Ikram, A., Saeed, F., Feng, X., & Chen, Y. (2023). Rabbit meat—Production, consumption and consumers' attitudes and behavior. *Sustainability, 15*(3). https://doi.org/10.3390/su15032008
Uddin, J., McNeill, D.M., & Phillips, C.J.C. (2023). Measuring emotions in dairy cows: Relationships between infrared temperature of key body parts, lateralised behaviour and milk production. *Applied Animal Behaviour Science, 269*, 106115, https://doi.org/10.1016/j.applanim.2023.106115
United Nations Food and Agriculture Organisation (2023). *Crops and livestock products.* https://www.fao.org/faostat/en/#data/QCL
Vaarst, M., Hellec, F., Verwer, C., Johanssen, J.R.E., & Sorheim, K. (2020). Cow calf contact in dairy herds viewed from the perspectives of calves, cows, humans and the farming system. Farmers' perceptions and experiences related to dam-rearing systems. *Journal of Sustainable and Organic Agricultural Systems, 70*, 49–57. https://doi.org/10.3220/LBF1596195636000
Vale, P., Gibbs, H., Vale, R., Christie, M., Florence, E., Munger, J. & Sabaini, D. (2019). The expansion of intensive beef farming to the Brazilian Amazon. *Global Environmental Change, 57*. https://doi.org/10.1016/j.gloenvcha.2019.05.006; https://s100.copyright.com/AppDispatchServlet?publisherName=ELS&contentID=S0959378018312093&orderBeanReset=true
Valpey, K.R. (2020). *Cow care in Hindu animal ethics.* Springer Nature. https://doi.org/10.1007/978-3-030-28408-4_1
Vets for Compassion (2024). *South Yarra cattle truck crash.* https://vetsforcompassion.org/2024/07/28/south-yarra-cattle-truck-crash/
Weary, D. M. & Chua, B. (2000). Effects of early separation on the dairy cow and calf: 1. Separation at 6h, 1 day and 4 days after birth. *Applied Animal Behaviour Science, 69*, 177–188. https://doi.org/10.1016/S0168-1591(00)00128-3

4

Pigs: Complex and intelligent individuals

INTRODUCTION

A Greek island and a small open pick-up truck set the scene for a tiny piglet's great escape. At just six weeks of age, the piglet took life in her hands and leapt from the back of the truck as it rattled across the island. Whether the piglet knew she was on her way to the slaughterhouse is unknown, but that fateful decision ensured that this little piglet's life story was only just beginning. But first there was a bicycle ride and a ferry trip before the piglet who would become known as Gucci found a home where she could express her unique personality in ways less dangerous than leaping from a moving truck. Similar to human newborns, piglets have no period when their stress-related adrenocorticotropic hormone (ACTH) receptor response is not functioning (Lind et al., 2007), with the hypothalamic-pituitary-adrenal (HPA) axis activated by a range of stressors including transport stress and psychological stress. The tiny piglet's sibling chose not to jump at that moment, possibly responding differently to the stressors they shared, and so the siblings' life trajectories deviated dramatically and irrevocably.

The pig's long and entwined history with humanity is unfamiliar to many urban dwellers whose sole contact with these charismatic creatures may be through the foods they eat, whether it be crispy bacon, ham or roast pork with crackling, or the gelatine in wine, chewing gum and jelly (Spry-Marqués, 2017). This contact may also extend to many less obvious products, which have made pigs one of the most utilised of all domesticated animals. Their strong bristly hair is used for brushes, their hide is used for products ranging from shoes to purses, pig bones contribute to some glues, their blood and offal are used in fertilisers, soaps and medical supplies, while many pharmaceuticals have been tested on pigs, along with tattoo artistry and dermatology. Pigs have even been genetically modified to allow their organs to be transplanted into human bodies.

The commodification of the pig contrasts sharply with increasing evidence that their intelligence is equal to, or even supersedes, that of companion animal dogs who are protected and idolised in many cultures. When not confined in breeding

 DOI: 10.1201/9781003518853-5

crates and pens, pigs form complex social networks and avoid soiling the areas in which they eat and sleep. Rolling in water and mud is essential to maintain coolness because, contrary to popular descriptors of "sweating like a pig", they cannot sweat at all. In many countries, pigs remain confined, forcing them to fulfil the stereotype of filth and obesity as they are selectively bred to grow ever larger until they are crippled under their own weight and unable to escape their faeces and vomit. They have also been associated with zoonotic diseases, with the world pandemic of 2009 labelled Swine Flu, and the subsequent African Swine Fever of 2018 resulting in the death of an estimated 25 per cent of the world's pig population (Long, 2019). Ironically, rather than lamenting the death sentence imposed on these sentient animals, it was largely reported in terms of the significant impact on the global protein market. Even when pigs escape human confines and establish wild populations, they are then labelled damaging pest animals to be culled.

It is paradoxical how an animal as smart as, or even smarter than, the domestic dog can be commodified, shunned by some individuals and whole cultures for their perceived gluttony and unclean ways, and used as a derogatory metaphor for a "pig-headed" human who is stupid, obstinate and set in their ways. Speciesism, the practice of seeing one's species as morally superior to members of other species and finding ways to justify that belief is endemic in human society (Ryder, 2011). Supporters of speciesism position nonhuman animals as inferior to humanity and therefore with less moral standing. This is particularly evident among low-status animals such as food animals and disgust-provoking animals that fail to trigger a person's nurturing or emotional response. No wonder the pig occupies a contested space in human society.

Before examining this contested space, Gucci's story shows what can happen when a pig takes a leap of faith into the arms of people who see her as an individual, rather than a commodity.

Gucci's great escape—Vrouva Farm, Greece

On the small Greek island of Aegina, Gucci, the escapee pig, lives her best life at Vrouva Farm. Thanks to an international community of committed animal lovers, some of whom live in Aegina and help out at the farm, Vrouva grew from strength to strength as a haven for animals rescued from neglect, starvation and abuse (Vrouva Farm, n.d.). Gucci the pig is one of those animals whose life trajectory was changed when she leapt from that truck to her forever home at Vrouva. As with all farmed animal sanctuaries, financial struggles are an ongoing issue, but climate change has introduced additional burdens in this tiny island retreat. Rising temperatures and the risk of fire in the surrounding pine forest are a testament to the changing world and emphasises the urgent need for humans to rethink their relationship with all aspects of nature. Vrouva and residents offer a compassionate example of how to live in harmony with the natural world and combat the speciesism, which sees sentient individuals such as Gucci sent off to slaughter before her life had even begun. But back to the amazing adventures of Gucci, as recounted by Vrouva founder Alexia Pap.

In 2020, a tiny six-week-old piglet and her sibling were travelling to the slaughterhouse in the back of an open pick-up truck when one of the piglets made the decision to leap for her life. While the maturation of the postnatal pig brain is in some ways comparable to that of humans, this feat far exceeded anything a human infant could do at that age. Taking her life in her hands, the little piglet jumped off the truck into the path of a woman on a bicycle who was coming from behind. From that moment forward, the piglet's life changed forever. Grabbing the tiny piglet and tucking her under her blouse, the woman headed home to work out what to do with the brave escapee. Contacting Vrouva Farm to find a future permanent home was the easy part. Vrouva was on a different island and only dead pigs—as pork meat—were allowed on board the boat that travelled between islands. Enter an experienced animal activist volunteer who placed the intrepid young escapee in a dog carrier, covered her up and warned anyone approaching that the dog within was aggressive. That stopped all nosy officious humans in their tracks.

Arriving at Vrouva was the beginning of a new life for the piglet who would become known as Gucci. For Gucci, speciesism does not exist. She loves sheep, Hope the cow, humans and Suzy, the other pig who was also saved from the slaughterhouse and is constantly playing pranks. Everyone loves Gucci back, except Suzy, who seems jealous of Gucci's popularity and good looks. The occasional nip puts Gucci in her place but breaks her heart as she cannot understand why someone would not like her. Of course, Gucci does not articulate this, but her body language conveys confusion and sadness that is hard to ignore or misinterpret, just as we would have no difficulty interpreting a child's body language when rejected by another child. When two cute lambs joined the Vrouva family, Gucci was perplexed at how someone else could be getting the attention—and snacks—that rightfully belonged to her. A quick nip to assert her cuteness dominance was never repeated, and life for Gucci returned to normal.

Gucci's huge character, intelligence and sense of humour and mischief are a constant reminder that pigs are unique sentient individuals. Instead of being slaughtered at six weeks, Gucci learned to play, something young pigs love to do. A toss of the head, scampering, jumping, pawing, pivots, flops and gambolling, Gucci has the space and freedom to enjoy it all in the safe haven of Vrouva Farm.

SOME BACKGROUND HISTORY

The exact timeline of the human/pig relationship remains open to debate, but there is consensus that it has been a long and varied journey. For millennia, wild boars were important prey animals for early hunter-gatherers across Eurasia, but it was during the Holocene (the last 11,700 years of the Earth's history) that this predator–prey relationship morphed into something that would have far-reaching implications for the domestic pigs of today. The shift from hunting-gathering to agriculture, and the domestication of animals and plants changed the life trajectory of pigs forever.

Figure 4.1 Gucci, the piglet who jumped from a moving truck to start a new life. (Image supplied by Vrouva Farm and reproduced with their permission.)

Moving backwards, the evolutionary story of wild boar species (*Sus scrofa*) and domestic pig subspecies (*Sus scrofa domesticus*), both members of the Suidae family, began almost 40 million years ago in Central Europe, followed by Africa and Asia around 20 or 25 million years ago (Spry-Marqués, 2017). Fast forward to the past one million years and there is evidence of *Sus scrofa* populations spreading out across almost the entire Eurasian continent, with relentless domestication being the inevitable outcome as human needs changed (Zhang et al., 2022).

Domestication of pigs is described as a critical event in the history of human agricultural civilisation, occurring in at least two locations—Anatolia (Near East where Asia and Europe meet) approximately ten thousand years ago and China approximately eight thousand years ago. Analysis of the genomes of modern pigs and boars suggests that the forebearers of European pigs interbred with at least two different populations of wild boars ranging between Europe and Anatolia, and pigs from East Asia similarly interbred with local boars (Frantz et al., 2015). Wild boars' ferocity was recognised and feared throughout human history, but domestication saw pigs kept in poor conditions, deprived of their biological needs and forced to adapt to human care and control through environmental manipulations and selective developmental changes (Maes et al., 2020).

Pigs are social animals who favour polygamy (Zhang et al., 2022). While wild boars breed once a year resulting in around five piglets, domesticated pigs can breed every five months, with double that number of piglets. As time passed, breeding sows were genetically selected and manipulated to reach weights three times that of their forebears in the Middle Ages, a necessary development to sustain global demand. Pork meat accounts for more than one-quarter of total protein consumed worldwide (Maes et al., 2020), making it the second highest per capita consumption of meat behind poultry and ahead of beef and sheep (Siddiqui et al., 2023). Intensive pig farming became essential to deliver high biological and economic productivity with a low input of labour, feed and space per animal.

PIGS IN CONTEMPORARY SOCIETY

In contemporary society, pigs hold the unenviable status among mammals of being the only livestock animal bred solely for slaughter. Their living body offers nothing in the way of wool, milk, eggs or labour, although they are increasingly being used for spare body parts for humans. Xenotransplantation requires the genetic modification of an animal to provide compatible spare parts for human transplantation. While selectively breeding a pig-human raised considerable debate around consent of human recipients and the potential transmission of disease, omitted from these ethical debates was the implication of breeding a pig whose sole purpose in life was to provide spare parts for the dominant species, humanity. Responding to the purported "medical break-through" in 2022, when a US man became the first human to receive a heart transplant from a pig–human hybrid, Julia Baines, science policy adviser for People for the Ethical Treatment of Animals (PETA; United Kingdom), commented:

> ...using animals as warehouses for spare parts is morally wrong. Pigs are sentient, complex, intelligent individuals that, like any animal, shouldn't be raised and raided for their organs. People who need a transplant need a human organ....
>
> *(Baines, 2022)*

Humans have taken so much from pigs, but not just through xenotransplantation. Over time and geographic location, pig farming changed to meet growing global demand. From the traditional smallholder rearing systems prior to World War 2 (Tzanidakis et al., 2021), which are still to be found in some developing nations, to extensive and intensive commercialised facilities in the second half of the twentieth century, pigs became a means to provide cheap food for growing global populations (Maes et al. 2020). The "megafarms" of the 1990s saw tens of thousands of breeding sows in a single location and accounted for more than half the global pig production. These settings sought to capitalise on mechanisms for feed efficiency, health management, environmental control, reproduction management, genetic selection for better performance and consistency of product for delivery to the insatiable market. Genetic selection of more prolific sows increased litter size, but this was accompanied by negative issues such as lower birth weight, longer farrowing duration, less colostrum and more competition between piglets. Sows weighing more than 315 kilograms (700 pounds) increased the crush risk when litters of 10 or 12 piglets weighing one kilogram (two pounds) each were born. This was used as a flawed rationale for confining sows to farrowing crates and further curtailing natural behaviours. Intensive farming also had implications for weaning time, dropping from a gradual process at 17 weeks of age in natural weaning to a more abrupt weaning at four or five weeks in intensive housing conditions.

With these megafarms also came increased risks of disease outbreaks and the development of stereotypies, the stereotypic behaviours that indicated all was not well in the pig's central nervous system and that pathological changes in the brain were underway (Tzanidakis et al., 2021). Despite increased body size crammed into reduced space, the sow's need to express natural behaviours such as exploring, foraging and instinctual nesting activities could not be quashed by generations of artificial selection. Intense restlessness prior to giving birth, biting the bars of the stall and obsessing over any straw or loose material in futile attempts to "nest" caused relentless frustration, stress and an ongoing welfare issue in industrial farming environments (Fraser & Weary, 2022).

The stereotypies triggered by chronic stressors such as confinement, restraint, social isolation or social defeat were also associated with lateralised hippocampal volume changes, similar to the changes observed in humans with stress-related conditions such as post-traumatic stress disorder (Lind et al., 2007). Assessing stress in pigs largely focused on aversive conditions affecting the autonomic nervous system and the HPA axis and measured by electrodermal (skin conductance) and cardiovascular (heart rate, blood pressure, heart variability) changes. Just like humans, cortisol, the colloquially named "stress hormone", is the primary active hormone in the HPA axis. The ACTH controls the synthesis of cortisol and is itself synthesised in the anterior pituitary gland (Murphy et al., 2021). Piglets, like human infants, have no period where the stress-related ACTH response is inactive.

Frustration, social isolation and, conversely, forced socialisation with unfamiliar conspecifics are some of the psychological stressors shown to activate a pig's HPA axis. Piglets are particularly vulnerable to psychological stress, which may occur through early weaning, separation from the sow, social isolation, different pen mates—or, in Gucci's case, an early trip to the slaughterhouse—resulting in ongoing changes in behavioural, neuroendocrine and immunological regulation and a flow-on effect to memory impairment (Lind et al., 2007). This may be compounded by a lack of environmental stimulation in intensive farm settings, thus limiting the choice of activities and even movements (Maes et al. 2020). Group formations, social hierarchies, long-term mother–offspring interactions and escape from aggression combine to frustrate a young pig's naturally curious and intelligent mind. They will never experience the natural behaviours of rooting, wallowing, exploring and playing, all vital to emotional wellbeing.

Play is fundamental to the physical, psychological, cognitive and social development of mammals, and pigs are no exception. Concentrated in the neonatal period, play is positioned as training for the future without which the development of stress coping mechanisms, social skills and aggression management skills are at risk of unsatisfactory completion (Martin et al., 2015). When exploratory and play behaviour is unfulfilled, increased fighting and frustrated activities such as painful tail biting have been observed.

Of the estimated 750 million pigs worldwide, almost three-quarters reside in Asia where pork is the leading product in the meat industry (Wang & Li, 2024). The outbreak and spread of African swine fever in 2018 caused pig populations to drop dramatically, compounded by the COVID-19 situation in 2020. However, pig slaughter and pork productions steadily increased after that, largely clustered around China, Republic of Korea and Japan in East Asia, as well as Vietnam, the Philippines, Thailand and Indonesia in Southeast Asia. China remains the leading pig producer in Asia aided by the development of a number of multi-storey pig farms capable of housing hundreds of thousands of pigs monitored by hi-tech cameras and fed via conveyor belts, further alienating pigs from their natural environments and behaviours (Andrews, 2023).

Despite these mega-farms and the infiltration of other medium-scale producers, a large population of pigs in Asia and Africa can still be found in traditional, small production systems—either backyards or small-scale commercial farmers who remain unmonitored and therefore at greater risk of harbouring infectious disease outbreaks coming out of the intensification and globalisation of pig production (Wang & Li, 2024).

In South Africa, it is estimated that around half of the more than one million pigs live in informal piggeries (Cowling, 2024), at risk of poor housing and sanitation, limited access to balanced food, water and veterinary care, and burdened with lung lesions, abscesses, ulcers, heavy parasite loads and a range of other infectious diseases. Their waste disposal pollutes water sources and soil, leaving a long-term impact on surrounding ecosystems in these impoverished communities. When the time comes, pigs are slaughtered by whatever means available, including stabbing and strangulation, in front of other pigs and humans alike.

The story of Peanut Butter and her piglets demonstrates that the physical and psychological harm experienced by pigs trapped in these informal piggeries can be just as damaging as that experienced in mega-farms. Importantly, Peanut Butter's story shows that both physical and psychological scars can be healed when relocating to a farmed animal sanctuary such as Greyton Farm Animal Sanctuary in South Africa.

Peanut Butter was part of Greyton's innovative "Pigs to Plants" programme offering informal pig farmers the opportunity to transition from farming pigs in often inhumane conditions to farming plants (Greyton Farm Animal Sanctuary, 2025b). Recognising that the pigs' welfare was inseparable from the poverty and cultural practices of humans, Greyton provided an equitable solution for both. In return for surrendering Peanut Butter to Greyton, the sanctuary supported the farmer to turn their pig space into a market garden in the heart of an impoverished community that so desperately needed food security.

Peanut Butter's story shows what the future can hold for a mother and her piglets when they find themselves in the safety of a farmed animal sanctuary.

Peanut Butter and her growing family—Greyton Farm Animal Sanctuary, South Africa

Greyton Farm Animal Sanctuary started with a simple dream to develop land sustainably and live in harmony with the natural world (Greyton Farm Animal Sanctuary, 2025a). In 2003, that dream became reality when Nicky Vernon, a British ex pat, purchased 40 hectares (99 acres) of land near Greyton on the Western Cape of South Africa and, with friend Rohan, created their own personal sanctuary in a farmhouse and other buildings made of straw bale and clay. But that was only the beginning of a sequence of events that turned a personal sanctuary into an animal rescue haven and provided the backdrop for Peanut Butter's new life.

First, there were the weak lambs, born in the surrounding fields and left to die. Their plight proved irresistible to the caring Nicky and Rohan who collected them and nursed them back to health. Then, there was Bella the pig's arrival from Cape Town. and the subsequent influx of "mini" pigs whose shocked owners discovered that their tiny wriggly pink bundles were anything but mini as their bodies continued to grow. And finally, there was the newborn calf Heaven whose mother had been dragged bellowing from the field and her fragile sickly calf left to die. Heaven did die later that same night, but peacefully with a warm full belly and surrounded by love, thus cementing Nicky's resolve to do all she could to stop such needless suffering. In 2014, Greyton Farm Animal Sanctuary was officially established to offer a forever home to abused, neglected, traumatised, sick, disabled or retired farm animals, ensuring that they could live in peace for the rest of their lives. With around 200 rescue animals, the days are long, and the work is constant for Nicky and the team, but a prominent quote on the Greyton Farm Sanctuary website says it all:

> "I get more joy, in one minute, from the rescued farm animals at the sanctuary than I ever did in decades of eating meat" (Nicola Vernon).

But back to the story of Mama Peanut Butter, as shared by Greyton founder Nicky Vernon.

Greyton has a Pigs to Plants project where backyard farmers keeping their pigs in awful conditions are asked to surrender their pigs into the care of the sanctuary and, in return, they are provided with an income generating vegetable garden.

The surrender of a large brown mamma pig under this programme added another 17 animals to the sanctuary. There were eight of the mamma pig's piglets, as well as eight piglets who joined the family following the slaughter of mamma pig's sister two weeks earlier. Pigs are not only intelligent caring creatures, but they are fiercely defensive and devoted mothers. It was no wonder that the mamma pig adopted her sister's piglets into her large brood.

On the mamma pig's arrival at the sanctuary, she was given the name Peanut Butter, thus establishing her as a unique individual and no longer a nameless commodity to be utilised for her body and breeding capacity. Peanut Butter had witnessed her sister's slaughter in the confined quarters of the backyard piggery and was understandably frightened of people. With limited agency over her social environment in barren housing, Peanut Butter's fear response would have been heightened by this event through a process called emotional contagion.

Pigs, like many other animals, have the capacity to experience emotional contagion, a rudimentary form of empathy. Defined as the arousal of emotion in one individual when witnessing the same emotion in another individual (Herrando & Constantinides, 2021), it was inevitable that witnessing and feeling her sister's slaughter would leave psychological scars on Peanut Butter. This was compounded by pigs' excellent long-term memory capacity (Lucas et al., 2024), resulting in a traumatised and mistrustful new arrival at Greyton along with 16 traumatised piglets affected by their mother's distress.

But pigs also have the capacity to emotionally heal when feeling safe and secure. Allowing Peanut Butter the space to forage, feel freedom for the first time in her life and avoid highly fearful stimuli worked wonders for the mamma pig and her offspring. Taking the piglets to be sterilised briefly returned her to a place of insecurity and trauma, but their safe return restored Peanut Butter's equanimity—especially after she gave birth to eight more piglets one month after arrival at the sanctuary. For the first time in her life, Peanut Butter could engage in pigs' natural pre- and postpartum behaviours forbidden to her in the confines of the backyard piggery.

Peanut Butter's complex and unique personality gradually evolved as she became a wonderful mother yet again and a confident, trusting and happy resident at the sanctuary. Importantly, the piglets finally had the freedom to engage in the serious business of play, as they scampered, jumped, flopped, learned to forage and enjoyed the rough and tumble of mock fights.

Nicky described the joy of living with Peanut Butter and her extended family: "She loves cuddles. Her older babies have grown more trusting, and her newest ones are completely happy with us. The best thing is going out into the fields with them, lying down and being smothered by hordes of happy, affectionate piglets and their doting mom".

Figure 4.2 Peanut Butter relaxing with some of her piglets. (Image supplied by Greyton Farm Animal Sanctuary and reproduced with their permission.)

FEELINGS AND EMOTIONS

Selective breeding for increased body weight gradually saw a corresponding reduction in the pig's absolute brain size and neocortical neurons (Minervini et al., 2016). The weight of the adult pig brain averages around 180 grams (0.4 pounds) with up to 400 million neocortical neurons, depending on body size, breed and duration of domestication (Cozzi et al., 2020). In comparison, the average weight of an adult human brain is around 1,300–1,400 grams (2.9–3.1 pounds) with 19–23 billion neurons (Lind et al., 2007). Despite these differences, the rate of development in the neonatal pig brain in many ways models the development of the human brain, making pigs a standard experimental subject for human embryology and brain development.

The similarities between the pig brain cortical surface and the folding of the human gyrencephalic neocortex, as well as similarities in the limbic structure (associated with emotion, motivation, memory and behaviour regulation) and the hippocampus (associated with learning and memory) further primed the pig as a chosen experimental model to avoid using the more ethically challenging and controversial primate brain (Lind et al., 2007). There were differences, of course, especially in areas such as encephalisation (the amount of brain mass relative to body mass) and the organisation of the main cortical lobes, but for researchers, similarities outweighed differences. As animal behaviourist Temple Grandin succinctly commented: "If the nervous system of a rat was totally different from that of humans, this research would be totally useless" (2018, p. 16), a sentiment equally applicable to the pig, whose brain became a sought-after commodity for researchers and the medical profession.

These similarities confirm that pigs, like most mammals, have the neural substrates conducive to experiencing mental states that are valenced—that is, experiences that are positive or negative, pleasant or unpleasant, rewarding or punishing. Pigs are capable of showing preferences, as identified in an early study where pigs actively chose ramps with shallow slopes and closely spaced horizontal footholds to walk up (Fraser & Weary, 2022), while on a small Greek island, Gucci preferred life over the unknown when she jumped from an open truck, leaving her sibling behind.

Traditional behavioural tests to identify emotional responses in pigs focused on the removal of negative stimuli associated with fear and anxiety to satisfy animal welfare standards (Murphy et al., 2021). As welfare standards changed to include positive experiences, qualitative behaviour assessments instead targeted the behaviour of the whole pig as an expression of their inner wellbeing. However, inferring emotional status from a commodified animal, irrespective of neurological evidence that this was possible, attracted criticism of anthropomorphic projections or biased observations, depending on the observer's interests (that is, veterinarian, researcher, farmer, animal rights activist, vegan or non-vegan).

Ethologist Marc Bekoff (2007) commented that when studying animals, humans can only describe and explain their behaviour using words that make sense from a human-centric point of view. This made anthropomorphism a powerful tool when making a qualitative behaviour assessment. Carl Safina further exemplified the importance of recognising the capacity for emotions in animals when asking: if we recognise that an animal is hungry or thirsty, why cannot we also recognise that they are happy or joyful when acting that way? (Safina, 2015). Nevertheless, despite the indisputable similarities and functional equivalences between human and pig nervous systems, it is impossible to get inside their heads to check their thoughts and feelings. This leaves interpretation of overt behaviours such as play—or absence of play—as a necessary option.

While those running farmed animal sanctuaries can provide multiple examples of positive emotions in pigs, the development of reliable and valid measures of affective states remains an important part of animal welfare. It is not possible to directly study the conscious experience of emotion in animals (nor humans as it is important to remember that humans, after all, are animals), therefore relying on a mix of behavioural, cognitive and neurophysiological tools can offer some insight. Intuitively, the humans who interact with pigs on farmed animal sanctuaries can easily interpret the emotions of their residents by vocalisations, body posture, facial expressions, tail movements and ear posture, as evidenced by the many rescued pigs living at Mother's Heart Sanctuary in Indonesia.

Lucky, CeCe, MeiMei, Alok, Blackie MeiMei, Ellie, Piggu and Piggi, unique individuals—Mother's Heart Sanctuary, Indonesia

On the Indonesian island of Sumatra, a special animal sanctuary is home to a group of unique individuals. Mother's Heart Sanctuary commenced with a very different name: "My Home My Animal's Haven", which was a fitting title for the founders' goal of opening their home to animals in need. Their aim was simple: to show the world that the lives of animals are important and worth saving. Animals are not simply commodities for human use.

In 2023, the sanctuary became a registered non-profit organisation and the first and only farmed animal sanctuary in Indonesia. The new name of Mother's Heart Sanctuary was born, beautifully explained in this heartfelt Facebook and Instagram post:

> It is a self-evident fact that each and every farm animal is not only an individual capable of feeling pain loneliness, love, joy, friendship, curiosity as well as abject fear, terror and suffering as we ourselves would as humans if we were also raised by those with absolute power over us and massacred en masse for our flesh … but also that each and every farm animal every single one of them had a mother who was not allowed to care for them. Indeed, they could not defend themselves, much less protect their babies from evil. Let this sanctuary be the mother for the few animals saved and inspire us to stop harming the innocent.
>
> *[Facebook, October 31, 2023. Reproduced with permission from Mother's Heart Sanctuary]*

Since then, the eight pigs whose stories follow next, as well as over 200 other animals, have lived safely and happily under the care of founders, Loong and Sing. Loong and Sing kindly shared these stories to demonstrate how each resident's unique personality and emotional capacity becomes evident once they find their forever home at Mother's Heart Sanctuary.

In November 2016, Lucky was one of the first pigs to be adopted, thus avoiding his destiny of being sold by the local breeder for slaughter. The super-smart young piglet captured Sing's heart and quickly adapted to the unexpected circumstances that allowed him to express his innate and contagious happiness. The spoiled and demanding young Lucky soon established his likes and dislikes—including crackers with rice—as well as sleeping on Loong and Sing's bed until he was six months old. The eight-year-old grey and white pig continued to demonstrate a strong will and quirky nature, including going to his room and slamming the door after a disagreement with one of the dogs. Lucky's best friends, nine-year-old CeCe and MeiMei, were also saved from slaughter and provide testament to the individuality of pigs who are allowed to live rather than die as food for humans.

CeCe and MeiMei were adopted in March 2015. The two girls quickly showed a love for humans as they settled into the safety of the sanctuary. They remain always the first to greet any visiting humans—unless of course snoozing with Lucky, or CeCe is occupied in her much-loved mud bath.

Big boy Alok was intended for slaughter until Loong and Sing stepped in. Since then, Alok's loyalty and constant happiness have made him a joy to have at the sanctuary. But Alok also has a special talent: protector of the weak and peace maker, a role he takes very seriously. He might look scary to those who do not know and understand his protruding front incisor teeth and eccentric nature, but he is a gentle giant who, like Lucky, loves crackers with rice.

Young Blackie MeiMei was the last survivor among ten siblings, and for a moment in time was loved despite a horrific start to life in 2016. Found lying next to a garage, barely breathing, the tiny black pig had been dumped and left to a slow death. The smell of rotting corpses clung to her living body, mixed with the faeces covering her tender skin. Thousands of maggots infested her

Figure 4.3 CeCe loves humans, her best friend MeiMei and mud baths. (Image supplied by Mother's Heart Sanctuary and reproduced with their permission.)

Figure 4.4 Alok is a gentle giant, peacemaker and protector of the other pigs. (Image supplied by Mother's Heart Sanctuary and reproduced with their permission.)

eyes and legs as she lay totally paralysed. Despite this cruel abandonment, Blackie MeiMei remained a gentle loving soul with a very strong will to live. Transported to the safety of the sanctuary, Blackie MeiMei was cleaned up and fed porridge with blended fruit and vegetables by syringe then spoon. After a week, the maggots were gone, and Blackie MeiMei gained sufficient weight and strength to enjoy her daily hydrotherapy sessions. By the third week, she could stand and walk by herself, and her happiness was evident as she played, rooted and sunbathed for the first time in her life. When Blackie MeiMei died unexpectedly one month later, she was mourned as a unique individual rather than the discarded nameless commodity left to die in squalor. Blackie MeiMei had been loved and experienced freedom and happiness in her short life, something many pigs never achieve. The social media posts proved Blackie MeiMei had lived and been loved: "RIP Blackie MeiMei. You are always in our hearts. We always love you".

Ellie joined the sanctuary in 2019 at the age of five years. Mother to five piglets, Ellie arrived traumatised from torture and bearing marks of horrific abuse inflicted on her by humans. A scar on the right side of her face, close to her eyes, showed evidence of being poked by a machete. A deep open hole infested with over 100 maggots near her left eye was a testament to more recent torture before Loong and Sing rescued the whole family just days before slaughter. Despite all this, Ellie's strong personality coupled with her gentleness and patience meant she remained a fiercely devoted mother who refused to be defeated by the pain and suffering of human cruelty. It took almost a year to heal Ellie's physical wound, and during that time, she patiently cared for her five rapidly growing babies, Lucas, Lucy, Luna, LiLi and LuLu, oblivious to the gaping hole in her head. After four years of being loved as an individual, Ellie died in 2023.

In 2020, two slightly different adoptees arrived at the sanctuary. Piggu and Piggi, female wild boars, were brought in by some children after their mother was killed for meat by hunters. Some of their siblings escaped, but it was likely they could not survive alone. Faced with two helpless infants only days old and with their umbilical cords still attached, Loong and Sing willingly accepted these wild boars into their sanctuary. Ironically, as humans increasingly encroach on their homes, wild boars have become a nuisance and a pest in the land that was once their own. Piggu and Piggi's unique personalities soon became evident as the curious and active pair settled into sanctuary life. As infants the smaller Piggi would sleep on top of Piggu, a melding of spiky striped hair that gave them both immense comfort and never failed to melt the heart of observers.

Gentle, happy, loyal, playful, curious, devoted mother, peace maker—every pig is different if only humans take the time to see beyond what their bodies have to offer. And every pig at Mother's Heart is named, loved and grieved when they die. Some die far too early because of what humans have inflicted on them, and tears are shed. The remainder have their traumas soothed as a new life opens up for them.

Vocalisations

Pig vocalisations have provided an important and practical means to assess real-time health information, thus improving efficiency in sick pig evaluation and environmental regulation in large production systems (Hou et al., 2024). In contrast, researchers Avelyne Villain and colleagues (2023) explored pig vocalisations within the context of the human–pig relationship and the pig's overall welfare, including emotional state. Previous studies reported diverse pig vocalisations linked to their emotional states, a finding supported in Villain and colleague's study among 60 weaned female pigs. They concluded that a piglet's degree of familiarity with a human could reflect the way they vocalised in and out of their presence. Piglet grunts seemed to change in acoustic structure depending on spatial proximity to a familiar human who had provided additional care, likely linked to a positive and more intense emotional state than when further away. When philosopher Rene Descartes asserted in 1646 that no nonhuman animal was sentient because the only evidence of sentience lay in the use of language (Veit & Huebner, 2020), he clearly had not listened to a pig. Research has moved beyond the need for language to demonstrate sentience, but an animal's inability to articulate their subjective experiences of the world and their feelings can make it easier for some people to allay feelings of cognitive dissonance associated with the treatment of sentient beings in intensive animal agriculture.

As research evidence accumulates to support awareness and emotion in these sensitive animals, the story of Angel the saddleback pig provides some insight into emotional vocalisations after rescue and rehabilitation at Horse and Farm Animal Sanctuary in Wales.

Re-imagining the future—Angel and Horse and Farm Animal Sanctuary, Wales

What horrible future awaited an ex-breeding saddleback pig advertised for free on an internet site?

"She's a bit underweight", the pig farmer stated when contacted for more details, adding that he would be handing her over to the first person who turned up. The younger breeding pigs were bullying her off the food and out of the arks—and with nowhere else to go, she was sleeping outside in the horse muckheap in the middle of a cold Snowdonia (Wales) winter.

Death, whether by hypothermia, starvation or slaughter, was the inevitable outcome for this pig who was of no further use to her human owner. The race was on to raise the 450 pounds needed to transport her from Snowdonia and certain death to Pembroke and a new life at Horse and Farm Animal Sanctuary. Sanctuary founder Lisa and her supporters achieved the impossible, swiftly collecting the full amount to cover the pig's four-hour journey to safety.

Residents of Horse and Farm Animal Sanctuary are given a second chance as they live out their natural lives in the welcoming, safe surroundings of this Welsh sanctuary. Despite being free from abuse, neglect or imminent slaughter, some may never fully recover physically from the harms and selective breeding previously inflicted on them by humans. However, all residents have the opportunity to recover psychologically and emotionally as Lisa applies her 20 years' experience in helping and working with animals to mend the wrongs inflicted on them. Over the years, this vegan run sanctuary has provided safe haven to horses, goats, cows, sheep, ducks, chickens, turkeys and, of course, pigs.

Lisa's description of the heartbreaking animal who stepped off the trailer clearly showed she had been on borrowed time: "I've never seen a pig in such bad condition. She was covered in skin sores. I could see her ribs, spine and her hip bones all sticking out under this dry, flaky skin that had sunk to the shape of her skeleton. Her legs were urine stained where she'd slept on the muckheap, all of her sores were bleeding".

Despite the obvious signs of starvation, the pig wagged her tail and gently took food from Lisa's hand. This gentleness and trust became a hallmark of the damaged pig who never once tried to bite Lisa's hand during the lengthy painful healing of her sores and her psyche. Following Lisa into the stable, the pig settled into the bedding of deep straw for the night, only to have seemingly vanished the next morning. The little pig had wriggled her emaciated body into a burrow beneath the straw for the warmest, safest night she had experienced in a long time.

During the fundraiser to collect the transport costs, Lisa had described the pig as "sweet angel" and so the name stuck. For several weeks, Angel was fed five times a day, so she could gradually rebuild her body. The urine stains faded, and with daily cleaning, the sores began to heal. Upon meeting the other pigs, Angel's initial combative response was understandable after spending much of her life fighting for food and space, and so she remained solitary until some rescue piglets joined the sanctuary. Gentle, affectionate Angel had found her new family of choice.

In typical intelligent pig fashion, Angel learned her name in two days. She still enjoys being hand-fed on occasion as she looks up at Lisa and talks with a unique sound reserved for communication with her trusted human. For Lisa and Angel, shared vocalisations strengthened the bond built on trust and care, as did the cuddles and belly rubs in the sunshine where Angel happily exposed her vulnerable belly.

Angel was five when her breeding life ended and her value as a commodity ceased. It is impossible to tell whether she has forgotten those five years of abuse. With a lack of verbal communication skills, the possibility of episodic memory, defined as the recollection of personally experienced past events (Kouwenberg et al., 2009), in nonhuman animals has always been controversial. Should it be confirmed that farmed animals have memory of the "what/where/when" aspects of an event and therefore the capacity to worry about past and future events, this introduces another dimension to the ethical treatment of animals. Domestic pigs can remember the "what" when using social memory to recall past interactions with conspecifics and to avoid dominant pigs, and the "where" when using spatial memory to locate feeders. There is also some evidence that they can simultaneously remember the "when" of a past event through the recollection of context ("which"), thus meeting the criteria for episodic memory. Of course, Angel's memory of past abuses will never be tested now that she has found her forever home at Horse and Farm Animal Sanctuary.

ETHICAL IMPLICATIONS

As humans transitioned from the hunter/gatherer lifestyle requiring the death of wild boars to permanent settlements and agriculture, there would be no reprieve for the domestic pig who continued to face a death sentence at human hands. All parts of these intelligent sentient animals have been harvested by humans, including body parts in a bid to sustain human longevity. They have been genetically manipulated, experimented on and forced to live in squalid or crammed environments as they await inevitable and premature slaughter. Studies increasingly demonstrate levels of intelligence equal to or superseding that of companion dogs—or young children—creating an inevitable moral dilemma.

Along with the increased knowledge of how a pig perceives the world, including their brain structure and their needs, comes the fear of discovering that their emotional capacity, not just their cognitive complexity, is too high to ethically raise them for slaughter, research or spare parts (Cozzi et al., 2020). The responsibility of understanding them more to maintain welfare standards creates cognitive dissonance and the impossibility of hiding behind the notion that farmed animals lack understanding of what is happening to them.

Figure 4.5 Angel talks with a unique sound reserved just for Lisa, her trusted human. (Image supplied by Horse and Farm Animal Sanctuary and reproduced with their permission.)

REFERENCES

Andrews, M. (2023, March 27). The rise of mega pig farms in China. *DirectIndustry e-Magazine*. https://emag.directindustry.com/2023/03/27/the-rise-of-mega-pig-farms-in-china/

Baines, J. (2022, February 2). Don't treat animals as spare parts for people. *New Scientist* (3372). https://www.newscientist.com/letter/mg25333720-300-dont-treat-animals-as-spare-parts-for-people-1/

Bekoff, M. (2007). *The emotional lives of animals: A leading scientist explores animal joy, sorrow, and empathy–and why they matter*. New World Library.

Cowling, N. (2024). Pig population in South Africa 2010–2022. *Statista*. https://www.statista.com/statistics/1306708/population-of-pigs-in-south-africa/#statisticContainer

Cozzi, B., Bonfanti, L., Canali, E., & Minero, M. (2020). Brain waste: The neglect of animal Brains. *Frontiers in Neuroanatomy*, *14*, 573934. https://doi.org/10.3389/fnana.2020.573934

Frantz, L.A., Schraiber, J.G., Madsen, O., Megens, H.J., Cagan, A., Bosse, M., Paudel, Y., Crooijmans, R.P., Larson, G., & Groenen, M.A. (2015). Evidence of long-term gene flow and selection during domestication from analyses of Eurasian wild and domestic pig genomes. *Nature Genetics*, *47*(10), 1141–1148. https://doi.org/10.1038/ng.3394

Fraser, D. & Weary, D.M. (2022). Applied animal behavior and animal welfare. In J.J. Bolhuis, L.A. Giraldeau, & J.A. Hogan, *The behavior of animals*, 2nd edition*: Mechanisms, function, and evolution* (pp. 251–280). John Wiley & Sons, Inc.

Grandin, T. (2018). My reflections on understanding animal emotions for improving the life of animals in zoos. *Journal of Applied Animal Welfare Science*, *21*(sup1), 12–22. https://doi.org/10.1080/10888705.2018.1513843

Greyton Farm Animal Sanctuary (2025a). *About Greyton Farm Animal Sanctuary*. https://greytonfarmsanctuary.org/our-story/

Greyton Farm Animal Sanctuary (2025b). *Pigs to plants*. https://greytonfarmsanctuary.org/pigs-to-plants/

Herrando, C. & Constantinides, E. (2021) Emotional contagion: A brief overview and future directions. *Frontiers in Psychology*, *12*, 712606. https://doi.org/10.3389/fpsyg.2021.712606

Horse and Farm Animal Sanctuary (n.d.). *About us*. https://www.horseandfarmanimalsanctuary.co.uk/about-us

Hou, Y., Li, Q., Wang, Z., Liu, T., He, Y., Li, H., Ren, Z., Guo, X., Yang, G., Liu, Y., & Yu, L. (2024). Study on a pig vocalization classification method based on multi-feature fusion. *Sensors*, *24*, 313. https://doi.org/10.3390/s24020313

Kouwenberg, A., Walsh, C.J., Morgan, B.E., & Martin, G.M. (2009). Episodic-like memory in crossbred *Yucatan minipigs* (*Sus scrofa*). *Applied Animal Behaviour Science*, *117*, 165–172. https://doi.org/10.1016/j.applanim.2009.01.005

Lind, N.M., Moustgaard, A., Jelsing, J., Vajta, G., Cumming, P., & Hansen, A.K. (2007). The use of pigs in neuroscience: Modeling brain disorders. *Neuroscience & Biobehavioural Reviews*, *31*(5), 728–751. https://doi.org/10.1016/j.neubiorev.2007.02.003

Long, W. (2019, September 18). One quarter of the world's pigs killed by African swine fever as disease spreads to South Korea. *ABC News*. https://www.abc.net.au/news/rural/2019-09-18/one-quarter-of-worlds-pigs-killed-by-swine-fever/11524134

Lucas, M.E., Hemsworth, L.M., & Hemsworth, P.H. (2024). Review: Early life piglet experiences and impacts on immediate and longer-term adaptability. *Animal*, *18*(Supplement 1). https://doi.org/10.1016/j.animal.2023.100889

Maes, D.G.D., Dewulf, J., Piñeiro, C., Edwards, S., & Kyriazakis, I. (2020). A critical reflection on intensive pork production with an emphasis on animal health and welfare. *Journal of Animal Science, 98*(Suppl 1), S15–S26. https://doi.org/10.1093/jas/skz362

Martin, J.E., Ison, S.H., & Baxter, E.M. (2015). The influence of neonatal environment on piglet play behaviour and post-weaning social and cognitive development. *Applied Animal Behaviour Science, 163*, 69–79. https://doi.org/10.1016/j.applanim.2014.11.022

Minervini, S., Accogli, G., Pirone, A., Graïc, J.M., Cozzi, B., & Desantis, S. (2016). Brain mass and Encephalization Quotients in the domestic industrial pig (Sus scrofa). *PloS One, 11*(6), e0157378. https://doi.org/10.1371/journal.pone.0157378

Mother's Heart Sanctuary (2023, October 31). Introducing Mother's Heart Sanctuary. *Facebook*. https://www.facebook.com/photo?fbid=653205860315744&set=a.497831742519824

Murphy, E.M., Melotti, L., & Mendl, M.T. (2021). Assessing emotions in pigs: Determining negative and positive mental states. In S. Edwards (Ed.), *Understanding the behaviour and improving the welfare of pigs* (pp. 1–42). https://doi.org/10.19103/AS.2020.0081.13

Ryder, R.D. (2011). *Speciesism, painism and happiness: A morality for the twenty-first century*. Andrews UK Ltd.

Safina, C. (2015). *Beyond words: What animals think and feel*. Henry Holt and Company, LLC.

Siddiqui, S.A., Gerini, F., Ikram, A., Saeed, F., Feng, X., & Chen, Y. (2023). Rabbit meat—Production, consumption and consumers' attitudes and behavior. *Sustainability, 15*(3). https://doi.org/10.3390/su15032008

Spry-Marqués, P. (2017). *Pig/Pork: Archaeology, zoology and edibility*. Bloomsbury Sigma.

Tzanidakis, C., Simitzis, P., Arvanitis, K., & Panagakis, P. (2021). An overview of the current trends in precision pig farming technologies. *Livestock Science, 249*, 104530. ISSN 1871-1413. https://doi.org/10.1016/j.livsci.2021.104530

Veit, W. & Huebner, B. (2020) Drawing the boundaries of animal sentience. *Animal Sentience, 29*(13). https://doi.org/10.51291/2377-7478.1595

Villain, A.S., Guérin, C., & Tallet C. (2023). The use of pigs' vocalisation structure to assess the quality of human-pig relationship. *Peer Community Journal, 3*, e36. https://doi.org/10.24072/pci.animsci.100190

Vrouva Farm (n.d.). *Who we are*. https://www.vrouvafarm.org/who-we-are

Wang, L. & Li, D. (2024). Invited review – Current status, challenges and prospects for pig production in Asia. *Animal Bioscience, 37*(4), 742–754. https://doi.org/10.5713/ab.23.0303

Zhang, M., Yang, Q., Ai, H., & Huang, L. (2022). Revisiting the evolutionary history of pigs via *De Novo* mutation rate estimation in a three-generation pedigree. *Genomics Proteomics Bioinformatics, 20*(6), 1040–1052. https://doi.org/10.1016/j.gpb.2022.02.001

5

Sheep: Resilient and unique personalities

INTRODUCTION

Disturbing images of live export sheep, packed into container vessels and shipped around the world, occasionally disturb the moral equanimity of the general population. Innocent victims caught up in the economic and cultural needs of humans, these sheep (and other live animal exports) experience unimaginable stress and suffering. This is amplified when things go wrong, and a shipload of live sheep is stuck at sea in distressing conditions with no consideration of whether the cargo lives or dies. There may be conflict at the destination port, or suspected disease outbreak on board the ship, or lengthy delays in sweltering conditions as happened in 2021 when another ship blocked the Suez Canal, but the end result is the same: countless innocent lives lost in horrific conditions to provide humans with live animals to slaughter.

With no personhood and their death viewed in terms of monetary loss, the distress these sheep (and other animals) experience becomes overshadowed by government negotiations, insurance claims and business decisions. Graphic news and social media images show sheep crammed together, creating a sense of helplessness in some that is alleviated by turning away from these images. The psychological phenomenon of psychic numbing, a flawed protective measure that instils a sense of indifference and removal of personal responsibility when a disaster becomes too large to contemplate or remedy, releases many viewers from the need to act. Whether the sheep arrive without mishap or fall victim to some of the worst maritime welfare tragedies in history is only one part of the distress that started for these sheep well before they boarded the vessel. Some sheep do not survive the long truck journey from farms, others give birth on the trucks and their infants fall through the slats. Sheep are then herded on board ships and penned for weeks during the sea journey, only to be slaughtered by whatever means the cultural norms specify once they reach their destination.

In contrast, Beans, a sheep who is no different to those collectively described as live animal exports, is provided with the opportunity to live the good life in

DOI: 10.1201/9781003518853-6

a farmed animal sanctuary in New Zealand. As a newborn alone for the first 24 hours of life, Beans had the good fortune to run bleating towards a human who would go on to establish a sanctuary in the lamb's name. Beans grew up in the freedom of the sanctuary, allowing his positive outlook on life, his intelligence and the ability to outsmart the humans to flourish. While the label "sheep" is often used derogatively to describe humans with a flock mentality and no original thought, Beans and the friends who would later join him proved this stereotype to be false. The story of Beans, founding member of Saint Beans Animal Sanctuary for Lambs, appears later in this chapter.

SOME BACKGROUND HISTORY

Domestic sheep (*Ovis Aries*) are often described as an important source of meat, fibre and milk among the global livestock sector, as well as an important biomedical model (Clark et al., 2017). But they are so much more, and their history makes for fascinating reading. As one of the earliest domesticated animals, the current-day sheep originated from the wild Asiatic mouflon (*Ovis aries orientals*) as they spread out from Anatolia (Asia Minor) approximately 11,000 years ago (Li et al., 2021). The Asiatic Mouflon was dark coloured with a soft, woolly undercoat, large horns and a natural moult every year, providing an important food source when humans were hunter-gatherers. When hunter-gatherers started looking for places to settle and grow their own food, sheep became a valuable addition to this new agricultural way of life.

Human migration to establish farming societies in neighbouring regions including Europe, north Africa and central Asia saw the domestic sheep become an integral part of this transition across many regions (Yurtman et al., 2021). Over the years, zooarchaeological data from animal remains within archaeological sites, as well as genomic evidence, revealed high levels of genetic heterogeneity compared to some other domesticated animals such as dogs and cattle, supporting the notion that domestication of sheep involved multiple regions.

The domestication of sheep—often to the detriment of their wellbeing—proved to be a success story for humans for several reasons (Hinch, 2017). The diversity of usable products (wool, meat and milk), coupled with the sheep's ability to adapt to extremes of temperature and lack of fussiness in food requirements, made them easy to manage under a range of environmental conditions. Any perceived flightiness was actively bred out through the retention of animals that could be most easily habituated to human handling. Add in behavioural characteristics of a social structure that primed the herd to follow a strong leader (sheep, human or dog) and breed promiscuously in captivity to produce precocial young who matured quickly, and the fate of domesticated sheep was sealed. Sheep had an additional advantage in that they liked to remain in their own home range, a natural instinct that was encouraged once nomadic humans settled into farming societies.

Although originally hunted as a food source, the value of their wool became so popular that specialist breeding started to replace the primitive domesticated populations, spreading from the Middle East into Europe, Africa and Asia (Chessa et al., 2009). Global estimates indicate that there are around 1.3 billion domestic sheep (Worldostats, 2025), represented by around 2,000 different breeds and crossbreeds (Hinch, 2017).

When humans discovered the versatility of the sheep's wool, selective breeding for continually growing soft white fleeces became an inevitable manipulation in their history of domestication. This not only cemented their complete dependence on human care and control, but would have far-reaching implications for any sheep leaving the producer's control. The softer, woollier fleeces so prized in the textile markets did not naturally fall off these selectively bred sheep, instead requiring annual human intervention to avoid the sheep from becoming overburdened with an ever-growing thick matted coat.

In 2021, rescuers were shocked to discover a sheep who had been separated from his flock for many years and rendered immobile under the weight of his unshorn wool. Rescued from certain death and relocated to Edgar's Mission in Australia, the story of Baarack made global headlines.

Baarack's story, as provided by Edgar's Mission, Australia

Edgar's Mission, home to Baarack from 2021 until his death in 2023, is located on 62 hectares (153 acres) in Lancefield, Victoria, Australia. Overshadowed by the mighty Macedon Ranges, at any one time, over 550 formerly farmed animals call it home. For visitors, Edgar's Mission is a magical experience and wake-up call to the cruelty inflicted on farmed animals so humans can benefit from their bodies. As a hardworking team treats every animal with the care and respect deserving all sentient beings, long-term resident Ruby, the Kelpie who once faced death after "failing" as a working sheep dog, spent 16 years greeting chickens, goats, sheep, donkeys, humans—in fact everyone—with a cheeky sweep of her tail.

Established in 2003 and named in honour of the first resident, a pig called Edgar Alan Pig, Edgar's Mission is one of those special places where farmed animals can live, age, die and be grieved as unique individuals. No longer faceless commodities subject to human needs and greed, each animal has the freedom to experience the positive emotions not possible in intensive farming environments. As the life stories and quirky personality traits of resident animals reach a global audience via social media platforms, these animals become ambassadors of change, one human at a time, under the banner "If we could live happy and healthy lives without harming others, why wouldn't we?"

But back to the story of Baarack, the hardy sheep whose life trajectory took a dramatic turn when spotted by a human who chose not to ignore the suffering of this lost animal. In February 2021, at the height of a hot Australian summer, the sheep who would be named Baarack was seen struggling to move around in the Cobaw State Forest in Victoria. On the verge of death and burdened with what would later weigh-in at 35-plus kilograms (77 pounds) of woolly fleece, Baarack's body was a writhing mass of matted fleece, insects, painful grass seeds, twigs and leaves. Despite his wool growing longer and longer, eventually blocking his vision, this stoic sheep had somehow managed to survive on shoots of grass and water pooled in puddles, even as his ability to bend and stand upright diminished under the weight of unshorn wool.

Selective breeding for wool to be harvested for commercial purposes necessitates sheep requiring at least annual shearing for their welfare. Their ancestor, the wild Asiatic mouflon with a coarse coat and downy undercoat that responded to the seasons, had no such requirement. They grew a dense coat in cooler months and then naturally shed it in the warmer months. Beans, whose story appears later in this chapter, retained the ability to naturally shed, but not so Baarack whose welfare was thus dependent on human intervention.

Baarack achieved global fame as his unwieldy body was gently carried from the forest and the removal of years of overgrown wool began. Before and after shots of this superhero showed the transition to a mere shadow of himself once relieved of the matted mass of fleece, dead and alive bugs and vegetation. Despite the additional pain inflicted when removing the barbed grass seed that had caused a painful ulcer when wedged between his cornea and eyelid, and the many hours required to painstakingly pluck countless grass seeds that had burrowed their way into his flesh, Baarack bore no grudges to the humans performing these necessary tasks. He could finally graze freely in the open field with Chloe and Molly Brown who would become his best friends.

At Edgar's Mission, Baarack was always the first to approach people in the paddock and offer his toothy grin as he held eye contact. It was this desire to be up close and personal that allowed the early identification and diagnosis of skin cancer in his nostril, another legacy of sheep not being designed for the harsh conditions of the Australian summers. Anywhere on the body that carries skin without fleece—the nose, lips, ears and surrounds of the vulva and perineum when exposed through tail docking and mulesing—leaves ageing sheep vulnerable to the damaging ultraviolet rays that grow worse each year. Baarack's death at the end of 2023 created an outpouring of grief for the gentle sheep who had finally found sanctuary for the last years of his life. A list of Baarack's favourite things, posted on social media soon after taking up residency at Edgar's Mission, leaves the reader in no doubt that Baarack was a unique individual, capable of living a full and emotional life, once relieved of the suffering imposed by human manipulations.

Baarack's list of favourite things is now his legacy towards recognising the rights of all animals to be acknowledged as individuals and making the world a better place for all of them. These simple pleasures are a testament to both the care and safety provided by Edgar's Mission and Baarack's capacity to forgive that made his last years so happy.

In no particular order, Baarack loved: breakfast; friends, especially Miss Clementine; living at Edgar's Mission; not being weighed down by 35.4 kilograms of extra wool; making worldwide news; grazing in paddocks and being an ambassador for his kind—showing the world how resilient and brave sheep really are!

A year later, with a sense of déjà vu, Alex was rescued from a similar plight. Relieved of a massive 40 kilograms (88 pounds) of fleece, Alex also settled into sanctuary life at Edgar's Mission, a shadow of his former self. In an ironic twist of fate, in some parts of the world, there has been some reverse selective breeding to develop sheep who naturally lose their coats each year. Shearing sheep can become uneconomical when the cost of labour outweighs the value of the fleece, hence the bid to return to the natural status quo—although too late for Baarack, Alex and other sheep burdened by ever-growing fleeces.

Not only do sheep remains appear in archaeological digs, but they also have a special place in historical documents, with United Kingdom author John Smith's *Chronicon Rusticum-Commerciale, or Memoirs of Wool* (1747) attesting that:

Figure 5.1 Baarack arrived at Edgar's Mission with his matted fleece full of insects, painful grass seeds, twigs and leaves. (Image supplied by Edgar's Mission and reproduced with their permission.)

Figure 5.2 A much happier Baarack, relieved of his matted fleece and ready to be friends with everyone at Edgar's Mission. (Image supplied by Edgar's Mission and reproduced with their permission.)

"Sheep are one species of cattle, which God, in the first generation of things, commanded the earth to produce and sustain, with a power to increase and multiply after their kind". The memoir goes on to describe how all but seven perished in the flood, with one sacrificed for worship and the other six to go on and breed abundantly—"for the service of God and man, viz. their carcasses for sacrifice and food, their wool and skins for clothing Noah and his descendants" (Bischoff, 1842, p. 67).

Lambs and sheep became very popular for sacrifice, especially during the Iron Age (1200 to 550 BCE), when they were killed for anything ranging from thanks for a good harvest to fertility rituals. Some were buried with their human owners or segmented into joints of meat as food offerings for the afterlife.

Sacrifice continued throughout Europe well into the Middle Ages and continues around the world in some cultures. For Muslims, sheep and lambs are a popular sacrifice performed by professional butchers or laypeople during the annual Eid Al-Adha four-day celebration (Oktay et al., 2021). When researchers Oktay and colleagues (2021) explored animal slaughter during this period in Gaziantep, Türkiye, they identified an increase in injuries among inexperienced people choosing to perform the sacrifice themselves rather than professional butchers. Citing the need for further education and safety precautions to help reduce the number of injuries, there was no mention of the welfare impact on the sheep being slaughtered in this way.

The diversity of genotypes in over 2,000 breeds of domesticated sheep has seen them adapted to a wide range of environmental extremes ranging from the tropics to deserts to high latitudes and high rainfall areas (Hinch, 2017). Around 60 per cent of the world's sheep population are located in temperate weather zones of Europe, Asia, South America, Australia and New Zealand, with the remaining 40 per cent in semi-arid tropics of Africa, Asia, India, Middle East and the highlands of East Africa (Morris, 2017). Consumption of meat is quite low compared to beef, pork and chicken, and less than two per cent of milk is from sheep. Even wool production is declining compared to other textiles, resulting in a universal decline in sheep numbers globally in the last decade.

SHEEP IN CONTEMPORARY SOCIETY

While some countries still favour intensive or semi-intensive systems, based on all-day indoor or outdoor confinement or partial confinement with controlled grazing, extensive farming is the norm in a number of countries around the world (Simões et al., 2021). Unlike cows, pigs and poultry who are synonymous with confinement in intensive farming systems and associated with images of physical and psychological harm, sheep welfare may be overlooked when kept in extensive systems perceived by consumers to be more animal-friendly and sustainable. However, this is not always the case among the various types of extensive sheep production around the world that rely on pasture feeding and the animals remaining outdoors for much, if not all, of the year.

Extensive farming systems are typical in United Kingdom and southern hemisphere countries such as Australia, New Zealand, South Africa, Chile and Uruguay where monitoring, care and human intervention can vary considerably depending on the number of sheep and property size (Munoz et al., 2019). Sheep have opportunity for control in areas of their lives such as social interaction, grazing, exploration and maternal behaviours, suggesting satisfactory provision of criteria 1–4 of the five domains of animal welfare. However, another set of welfare concerns may arise depending on the geographic location of the farm and weather conditions, as well as the size of the farm and sheep-to-human labour ratio.

When comparing Scotland with a total of 2.6 million breeding ewes and an average flock size of 500 sheep and Australia with 37.2 million breeding ewes and average flock sizes of more than 2,800 sheep, there is a stark difference in sheep-to-labour ratios (Munoz et al., 2019). Welfare studies among large flocks report high neonatal lamb and ewe mortality, poor nutrition despite natural grazing, increased intestinal parasites, flystrike, mastitis and lameness and difficulties with provision of water and shelter. While sheep are yarded for husbandry procedures such as ultrasound pregnancy diagnosis and weaning, there remains a difficulty in identifying and managing individual welfare issues in large-scale extensive systems, especially when having to balance the economic value of one individual ewe or lamb against the monetary and labour costs of intervention. When combined with the distances that sheep may walk in extensive farming systems, lameness or congenital leg issues in lambs can be a significant welfare issue if not addressed early (Silva et al., 2022).

Extreme climatic conditions, including very high or low temperatures, droughts, snow, frost, wind exposure, solar radiation and humidity depending on the region, contribute to both poor quality pastures and sheep welfare in ways not experienced by intensively farmed animals. Each hemisphere has its own unique challenges, and inevitably, some newborn lambs get left behind and die. Sheep in Australia experience relentless drought and heat, while the cold, rain or snow of other regions can increase the risk of neonatal hypothermia and death. In New Zealand, home to almost 18 million breeding ewes (Munoz et al., 2019), one newborn lamb found himself alone and on the verge of hypothermia when his life trajectory changed dramatically. The story of Beans reveals how meeting and saving one lost lamb resulted in a whole new way of life for a man and the future lambs and sheep for whom he would provide safe sanctuary in New Zealand.

Beans, a timely intervention—Saint Beans Animal Sanctuary for Lambs, New Zealand

Who could predict that a tiny blob of white bounding up to a man taking a brief respite from the corporate world would lead to the creation of a lamb sanctuary in New Zealand? Saint Beans is New Zealand's first farmed animal sanctuary devoted to lambs and sheep, all of whom are loved and able to live out their natural life spans in safety. Newborn lambs are unable to regulate their body temperature, leaving them vulnerable to painful death from hypothermia. When born across the coldest wettest months, hypothermia becomes a real risk to lambs who have not received sufficient colostrum from their mother. For many reasons, including maternal death, rejection of the lamb after a difficult birth, poor maternal instincts or just unfortunate separation during a storm, some neonatal lambs are doomed to the intense drawn-out suffering leading to a miserable death by hypothermia. Lambs are born with brown fat around their heart and kidneys, which is their only source of energy in the first few hours of life. That is used up very quickly, and without colostrum, their bodies are deprived of energy and heat, resulting in hypothermia, potential organ shutdown and death (Lamb Care Australia, 2023).

Beans was found abandoned as a newborn, and while his enthusiastic approach to Brendon, the human who was taking a stroll on a beautiful winter's morning, showed he had not yet succumbed to hypothermia, it was clear he could not remain alone for much longer. But Beans was a survivor and a very noisy one at that. The odd white shape that jumped up and ran towards Brendon, bleating at a volume seemingly impossible for such a tiny creature, proved irresistible. Whether the lamb adopted Brendon or vice versa, he remained glued to the man's side as he intuitively recognised that his salvation had just walked over the hill. After the initial dash towards Brendon, it became clear that the lamb had been alone for over 24 hours and was weak, although remaining "full of beans" even before receiving his first life-saving drink of colostrum. A warm bed and full belly allowed the newly named Beans to begin his recovery with an irresistible love for life.

And there was so much to love about his new life. He slept on a lazy boy chair, spent considerable time running around the garden and hated being separated from Brendon even for a minute. That meant accompanying Brendon on trips and sleeping next to him while he worked, just as cats and dogs do in so many homes. Dogs, cats and lambs are all fundamentally the same, separated only by the engrained social and cultural norms that create an artificial divide between companion animals raised to become cherished human family members and lambs bred and valued for their body parts. However, once a sheep learns to trust, their behaviours can become very "dog like" as they race over excitedly to greet their human.

While Beans was rich in personality, clever and happy, he still needed at least one friend (other than Brendon). Sheep are herd animals, with a strong instinct never to be alone and the need to learn from each other. Enter Frosty, a lamb who had lost his mother and brother and was desperately in need of a friend. Within a day or two, Beans and Frosty had bonded, remaining inseparable best friends ever since. With the reassurance of a friend, Beans could move into the hay barn, and the two made themselves hay bunks from which they never deviated: Beans on the bottom bunk and Frosty on the top bunk.

As a Wiltshire sheep, Beans retained a natural shedding behaviour. He grew a thick coat in winter, then shed it in summer, unlike sheep such as Baarack who have been selectively bred to grow excess wool that they never shed. Beans' first coat shed left two perfectly symmetrical wool angel wings on his back, a fitting inspiration for naming the farmed animal sanctuary Saint Beans after the charismatic young lamb with a will to live.

Like all animals who are given the opportunity to reach their potential, Beans grew into a unique and affectionate individual with a relentlessly positive outlook on life. He remained close to Brendon, although frequently outsmarting him with his high intelligence and playful cheekiness. Beans knew his name as well as any dog, but that did not mean he would always respond. Instead, a turn of the head and an "okay, but what do you want?" look made it perfectly clear that Beans would respond on his own terms.

Farmed animals are often stereotyped as mindless and unfeeling, especially sheep whose name is used as a derogatory descriptor for mechanical following and stupidity. While this may make people who consume or wear sheep more at ease with their actions, nothing could be further from the truth about this complex and sensitive animal.

Figure 5.3 Adult Beans relaxing on the grass at the sanctuary that bears his name. (Image supplied by Saint Beans Animal Sanctuary for Lambs and reproduced with their permission.)

A convenient research model

Weighing in at approximately 130–140 grams (approximately 4.5–5 ounces), the brain of the sheep has been investigated within the context of how it can provide a convenient animal model for mapping human disorders and neurosurgical and neuropathological research (Cozzi et al., 2020). Because of its similar size and anatomy, sheep brains are popular models for specialities ranging from traumatic brain injury to neurological disorders (Trovatelli et al., 2022). Evolutionary history of mammals means many of the cellular and molecular mechanisms in humans and sheep are sufficiently similar to use sheep in investigations of the human nervous system and related disorders, as evidenced by the Huntington's Disease studies.

Huntington's Disease is a fatal hereditary disease resulting in relentless cell death in the brain's basal ganglia region. The possibility of using sheep to study the progression of this human condition arose in the early 1990s, and by 2010, the development of a Huntington's Disease transgenic sheep provided a popular model of prodromal or early-stage Huntington's Disease (Howland et al., 2020). The very traits that position sheep as unique individuals became valuable attributes in these studies. Sheep are long-lived, social animals, with the ability to recognise faces and with long memories. They learn quickly, allowing them to engage in experiments such as the cognitive tests similar to those given to humans and enabling observation of the gradual mental and motor decline synonymous with

Huntington's Disease. For sheep, being too expressive and intelligent led to their commodification as neurological research subjects.

Sheep brain studies are not new. Notes on the cortico-spinal or pyramid tract fibres of the sheep brain appear as far back as the 1870s, with much of the early studies focusing on electrical or mechanical stimulation of the different brain regions to map body response to stimuli and thus create an anatomic atlas of the sheep brain (Banstola & Reynolds, 2022). By 2002, Opdam and colleagues described the sheep (merino) model of focal epilepsy using concurrent EEG and fMRI to identify the localisation of coordinates for surgical placement and electrodes, although the researchers did not progress these findings to the structures of the human brain.

Other overlapping features of sheep and human brains include electroencephalographic elements, neuroradiological features, neurovascular structures and skull anatomy (depending on the breed) related to thickness, porosity and curvature of the skull (Trovatelli et al., 2022). Sheep and humans also have a similar gyrification index. This index describes the ratio of total neocortical surface area, including sulci (grooves in the cerebral cortex), to superficially exposed neocortical surface area. As comparison, mice have a score of 1.0, rabbits 1.2, cats 1.5, dogs 1.73, sheep 2.29 and humans 2.56 (Peruffo & Cozzi, 2014).

As often happens with farmed animals, sheep research proved more acceptable to animal ethics committees and the general public compared to companion animals and primates (Banstola & Reynolds, 2022). They were readily available, less expensive, and their larger brains allowed for greater relevance of translational research over large animal species compared to the rodents that are also commonly used in research. There were of course a range of differences, including weight disparities (140 grams [5 ounces] for the sheep brain compared to 1,400 grams [49 ounces] for the human brain) and a more developed cerebral cortex in the human for higher cognitive functions such as thinking, memory and language, but both sheep and humans shared the capacity to feel pain. Mammals in general have similar nociception, making it safe to assume that how pain is sensed and the physical and neurological processes behind this will be similar across sheep and humans. And yet there remains a lingering perception that it is ethically acceptable to inflict pain on some commodified animals reared for food compared to companion animals or humans. Farmed animals are also at risk of reduced pain relief depending on the estimated value of an individual animal to the production system (Steagall et al., 2021).

Pain perception

Imagine having flaps of skin cut from around your hindquarters to create an area of bare, scarred skin. Even with legislated pain relief for the procedure, there can be weeks of suffering as these bare patches heal and scar over. For over a century, many lambs in Australia have been subjected to this procedure called "mulesing" to prevent flystrike, the horrible situation where blowflies lay eggs in the folds of skin around the hindquarters and tail, and the hatching maggots feed off the sheep's flesh (RSPCA, 2025).

As prey animals, sheep have evolved to hide pain to reduce the risk of predation, but it is hard to ignore behavioural indicators such as reduced sociability, restlessness, decreased food consumption, abnormal vocalisations, altered gait, foot stamping, writhing and reluctance to move associated with these painful experiences. Sadly, the financial impact when immune function and quality of meat are compromised in animals experiencing chronic pain can attract more attention than the emotional effects of pain. Procedures such as mulesing, tail docking, castration, surgical artificial insemination or embryo transfer among ewes, and rough handling during shearing can create physical pain with a flow-on effect to the negative affective states of anxiety and depression. This can worsen pain perception and compromise overall welfare in these sentient beings (Steagall et al., 2021).

FEELINGS AND EMOTIONS

By 2001, it was clear that sheep possessed long-lasting face discrimination skills for both human faces (cross-species paradigm) and other sheep (Peirce et al., 2001). This skill, deemed so important in humans, was further supported and expanded in a 2017 study that concluded sheep have advanced face-recognition abilities similar to those of human and non-human primates (Knolle et al., 2017). Eight sheep learned to recognise the faces of unfamiliar individuals from photographs, even when the images were presented in different perspectives. When a photograph of their handler's face was included, sheep were also able to identify this familiar real-life person from a two-dimensional image, suggesting that sheep have holistic face-processing abilities, a skill jealously guarded by humans. An earlier study demonstrated that face pictures of familiar conspecifics presented to socially isolated sheep reduced levels of stress and fear (da Costa et al., 2004), reminiscent of the way humans find solace from photographs of loved ones. Electrophysiology studies showed that similar neural networks for face perception are activated in sheep, humans and monkeys, further narrowing the gap between humans and the animals they stereotype as anonymous mindless food sources.

Among the research of sheep's abilities to recognise faces, an additional observation was noted: sheep prefer to look at other sheep faces that are calm rather than frightened or startled faces (Kendrick et al., 2001). In fact, sheep will actively seek out positive emotional experiences, once they feel safe and secure. Part of this security comes from special friends, such as Beans and Frosties' immediate rapport, and the long-term memories embedded in the same neural mechanisms that humans use to maintain and recall memories of favourite friends over months or years.

Anyone who has observed lambs at play would have no doubt that these delightful animals are enjoying a positive emotional experience. While play is a crucial learning process for young animals, it also embodies uncensored emotional pleasure which is difficult to resist. Many farmed animal sanctuary workers describe the joy they feel even on the busiest of days when the lambs start leaping, head-butting and running around with their friends. This example of cross-species emotional contagion—the arousal of emotion in an individual when witnessing the same emotion in another individual—can be irresistible for humans and lambs alike. However, emotional contagion can negatively impact farmed animals when observing the abuse or slaughter of their conspecifics, or separation or loss of a group member.

The strong bond between ewe and her lamb is no different to so many other mammals, whether wild animals, companion animals or farmed animals. The bond begins within the first six hours of life, reinforced by olfactory cues that establish and lock-in recognition of the infant. Ewes with compromised olfactory senses draw on other senses, including auditory and visual cues, to maintain this crucial bond (Hinch, 2017), which links mother and offspring for at least two and a half years if given the opportunity.

A sheep mother's grief has been immortalised in the 1878 oil painting in the National Gallery of Victoria, Australia, painted by a little-known artist August Schenck and described as the saddest painting in the gallery. "Anguish", also known as the "sad sheep painting", stands 1.5 metres (5 feet) tall and more than 2.5 metres (8.2 feet) wide, and it was first shown at the Paris Salons. The painting portrays an ewe standing protectively over her deceased lamb in the snow, while a murder of crows collects around her, patiently awaiting a moment of weakness to descend on her dead lamb. The grieving mother's face shows determination and sorrow as her mouth is partially open in what the viewer imagines is a plaintive bleating for her lost lamb. The grim emotion has seen "Anguish" twice voted one of the gallery's most popular works, in 1906 and 2011 (Time Out, 2025), an ironic example of the cognitive dissonance that allows some humans to relate to the grief in this image on an emotional level while ignoring the paradox of consuming the flesh of this same animal.

The grief reflected in "Anguish" can be played out multiple times within the life of a farmed animal. Whether farmed animals experience grief is a contested notion, as to acknowledge the possibility means accepting responsibility for separating a mother from her newborn or ignoring the behavioural indicators that can eventually cause dispirited resignation and emotional shut down. For sheep in extensive farm settings, grief may be a luxury they cannot afford, as to linger over a dead lamb or friend risks losing the social support of the herd as they are moved on to new pastures.

Both intensive and extensive animal farming have made death and disability synonymous. For animals bred to physical extremes, confined in ways that contort bodies painfully and permanently, or bearing injuries unnoticed and untreated, disability goes hand in hand with a "better off dead" outcome. A simple cost–benefit analysis confirms the lack of economic viability to rehabilitate individual disabled animals who are likely headed for slaughter well before their natural lifespan.

However, this narrative can change for those animals who find safety in a farmed animal sanctuary. When Penny the sheep arrived at Strong Hearts Farm Sanctuary in Australia, she was very unwell. Pregnancy toxaemia, a metabolic disorder caused by low glucose concentrations in the blood and the breakdown of body fat to compensate, and a difficult birth had taken their toll on her vulnerable body. At Strong Hearts Farm Sanctuary, animal residents are treated with compassion and respect as they start their healing journey. Located in rural Victoria, Australia, the Strong Hearts team works to save and rehabilitate animals, while simultaneously educating and advocating to make the world a better place for all those who cannot speak for themselves. This is the story of Penny, the three-legged sheep who never believed in disability and whose emotional capacity was so strong that she could bend humans to her will.

Penny's journey—Strong Hearts Farm Sanctuary, Australia

Unfortunately, irreversible nerve damage and muscle contracture to one of Penny's legs increasingly hindered her ability to thrive. Penny was at risk of becoming disabled and was a target for the human assumptions and prejudices associated with ableism, which are often projected onto nonhuman animals. In May 2023, Penny officially became disabled when her right hind leg was amputated. This decision was made to relieve her unrelenting pain, but also in response to her actions, which made it clear that she retained a strong will to live and that nothing would hold her back.

When humans label an animal as disabled, they cannot know for sure how—or even if—that animal comprehends its physical or cognitive difference. Penny's vibrant and happy nature suggested she had no concept of what it meant to be different to four-legged sheep. The day after losing her leg Penny stepped outside into the sunshine, her spirit and will remaining fully intact after the marathon surgery. Within three days, she was walking more distance with less assistance and greater confidence, and within ten days, Penny was out in the world on three legs with no worries. Despite a prolonged secondary infection post-surgery, two months later Penny emerged more mobile than ever. She had quickly conquered getting around and laying down unaided, but once laying down, was unable to position her remaining hind leg under her body to stand up again.

Understanding human disability often focuses on the medical narrative where disability is a personal tragedy, with disabled persons requiring pity (Loseke & Green, 2019). While disability scholars have argued against this medical model of suffering among humans, it tends to linger in animal industries where the basic right to recognise suffering in animals was only achieved after a hard-fought battle (Taylor, 2020). Penny's inability to rise from a downed position could have put her at risk of perceived suffering and pity, but this sentiment did not appear in Penny's repertoire. With visits to the physiotherapist and more of her relentless determination, Penny once again demonstrated that disability was a social construct that definitely did not apply to her.

Penny could not articulate her joy in life, but observation of her facial expressions, behaviours and vocalisations left those who knew her best in no doubt of her capacity to have feelings of joy, affection, friendships, sharing, peacefulness, anticipation—or jealousy when other animals encroached on her treats or her humans. While this may raise accusations of anthropomorphism, ethologist Marc Bekoff (2007) commented that humans can only describe and explain an animal's behaviour using words that make sense from a human-centred perspective. Watching an animal, accepting differences and resisting the temptation to fall back on familiar interpretations embedded in the human experience—such as ableism—makes anthropomorphism a powerful tool to make sense of the thoughts and feelings of animals.

Penny's humans described her as having the most expressive face they had ever seen in a sheep, coupled with the attitude of a spoiled princess who reserved a special "baa" for when she needed or wanted something. The volume increased to ""MMMAAAAAAHHH" when asked her favourite question: "Do you want oaties?" Failure to deliver on the promised "oaties" in a timely way resulted in a look of sheer disappointment guaranteed to trigger guilt in her lax humans.

In an intensive or extensive farm setting, disabled animals are often presented as offering nothing back to their communities. This could not be further from the truth of Penny's life in the farmed animal sanctuary community. In a demonstration of interspecies emotional contagion, Penny's joy in the life that was handed back to her at Strong Hearts brought delight to all those in the vicinity of this three-legged sheep who had refused to believe in disability.

In March 2025, Penny passed away, surrounded by the people who had loved her and shared her journey at Strong Hearts Farm Sanctuary.

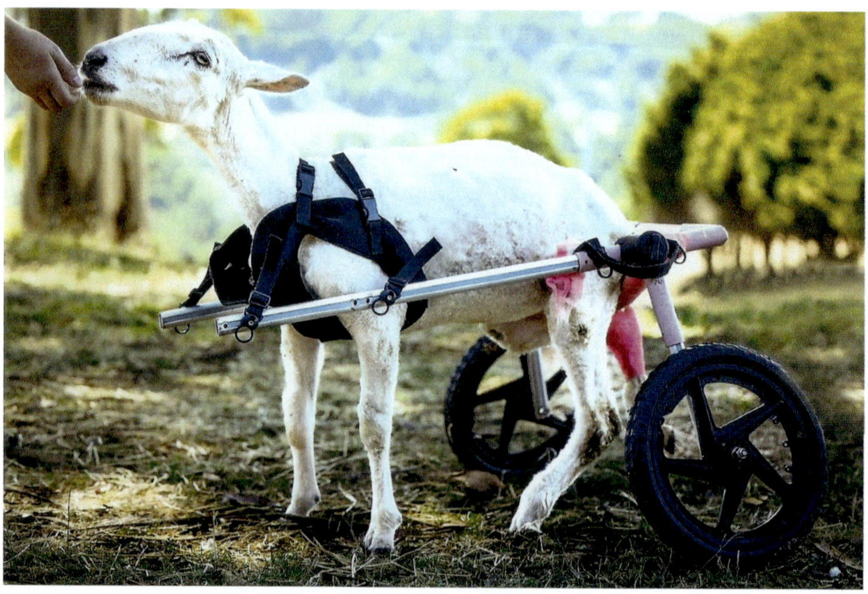

Figure 5.4 Penny's expressive face. (Image supplied by Strong Hearts Farm Sanctuary and reproduced with their permission.)

ETHICAL IMPLICATIONS

With mounting evidence of emotional capacity at all stages of a sheep's life, it is hard to understand why they continue to be perceived as food and wool commodities. While the long-term use of sheep brains in research inevitably led to a greater understanding of this complex animal, it also drew attention to the moral responsibility to know more about how sheep perceive the world when being used in this way. Not to do so raises the disquieting possibility that the main reason for failing to study the sheep brain for itself, rather than for what it can offer human research, may be fear of discovering that their level of cognitive and emotional complexity prohibits their ongoing use, abuse and slaughter for human benefit (Cozzi et al., 2020).

The fact that the sad sheep painting, "Anguish", is so popular illustrates the paradox which plagues humanity's relationship with food animals. This one mother and infant personalises a mother's fierce protectiveness of her infant, and yet thousands of lambs are born in dire circumstances such as transport, live animal export and extreme weather conditions. Many of these lambs are separated from their mothers, never receiving the opportunity to be protected in any way.

Sheep are more than empty-headed livestock who obediently follow, a stereotype reinforced when applying their name to humans perceived as unable to think for themselves. Their flocking instinct is grounded in the need for security and protection from predators, representing a common-sense survival response rather than stupidity. They have complex emotions and attachment bonds and demonstrate social skills and intelligence to rival many other animals. And yet humans have commodified their bodies for meat and wool, with no regard for the uniqueness of every lamb and sheep.

REFERENCES

Banstola, A. & Reynolds, J.N.J. (2022). Mapping sheep to human brain: The need for a sheep brain atlas. *Frontiers in Veterinary Science*, 9, 961413. https://doi.org/10.3389/fvets.2022.961413

Bekoff, M. (2007). *The emotional lives of animals: A leading scientist explores animal joy, sorrow, and empathy–and why they matter*. New World Library.

Bischoff, J. (1842). *A comprehensive history of the woollen and worsted manufactures and the natural and commercial history of sheep, from the earliest records to the present period*. Vol. 1. Smith, Elder.

Chessa, B., Pereira, F., Arnaud, F., Amorim, A., Goyache, F., Mainland, I., Kao, R. R., Pemberton, J. M., Beraldi, D., Stear, M. J., Alberti, A., Pittau, M., Iannuzzi, L., Banabazi, M. H., Kazwala, R. R., Zhang, Y., Arranz, J. J., Ali, B. A., Wang, Z., ... Palmarini, M. (2009). Revealing the History of Sheep Domestication Using Retrovirus Integrations. *Science*, *324*(5926), 532-536. https://doi.org/10.1126/science.1170587

Clark, E.L., Bush, S.J., McCulloch, M.E.B., Farquhar, I.L., Young, R., Lefevre, L., et al. (2017). A high resolution atlas of gene expression in the domestic sheep (*Ovis aries*). *PLoS Genetics, 13*(9). https://doi.org/10.1371/journal.pgen.1006997

Cozzi, B., Bonfanti, L., Canali, E., & Minero, M. (2020). Brain waste: The neglect of animal Brains. *Frontiers Neuroanatomy*, 14, 573934. https://doi.org/10.3389/fnana.2020.573934

da Costa, A.P., Leigh, A.E., Man, M.-S., & Kendrick, K.M. (2004). Face pictures reduce behavioural, autonomic, endocrine and neural indices of stress and fear in sheep. *Proceedings of Royal Society of London*, 271(1552), 2077–2084. https://doi.org/10.1098/rspb.2004.2831

Hinch, G.N. (2017). Understanding the natural behaviour of sheep. In D.M. Ferguson, C. Lee, & A. Fisher (Eds.), *Woodhead Publishing series in food science, technology and nutrition, advances in sheep welfare* (pp. 1–15). Woodhead Publishing. https://doi.org/10.1016/B978-0-08-100718-1.00001-7

Howland, D., Ellederova, Z., Aronin, N., Fernau, D., Gallagher, J., Taylor, A., Hennebold, J., Weiss, A. R., Gray-Edwards, H., & McBride, J. (2020). Large animal models of Huntington's Disease: What we have learned and where we need to go next. *Journal of Huntington's Disease, 9*(3), 201–216. https://doi.org/10.3233/JHD-200425

Kendrick, K.M., da Costa, A.P., Leigh, A.E., Hinton, M.R., & Peirce, J.W. (2001 Nov 8). Sheep don't forget a face. *Nature, 414*(6860), 165–1666. https://doi.org/10.1038/35102669. Erratum in: *Nature*. 2007 May 17; *447*(7142), 346. PMID: 11700543.

Knolle, F., Goncalves, R.P., & Morton, A.J. (2017). Sheep recognize familiar and unfamiliar human faces from two-dimensional images. *Royal Society Open Science*, 4, 171228. https://doi.org/10.1098/rsos.171228

Lamb Care Australia (2023). *Lamb Hypothermia*. https://lambcareaustralia.org.au/lamb-health-lamb-hypothermia/

Li, R., Yang, P., Li, M., Fang, W., Yue, X., Nanaei, H.A., Gan, S., Du, D., Cai, Y., Dai, X., & Yang, Q. (2021). A Hu sheep genome with the first ovine Y chromosome reveal introgression history after sheep domestication. *Science China Life Sciences*, 64, 1116–1130. https://doi.org/10.1007/s11427-020-1807-0

Loseke, D.R. & Green, S.E. (2019). Exploring narrative as a social science framework on disability and disabled people. In S.E. Green & D.R. Loseke (Eds.), *New narratives of disability: Constructions, clashes, and controversies* (pp. 1–8). Emerald Publishing Limited.

Morris, S.T. (2017). Overview of sheep production systems In D.M. Ferguson, C. Lee, & A. Fisher (Eds.), *Woodhead Publishing series in food science, technology and nutrition, advances in sheep welfare* (pp. 19–35). Woodhead Publishing. https://doi.org/10.1016/B978-0-08-100718-1.00002-9

Munoz, C.A., Campbell, A.J.D., Hemsworth, P.H., & Doyle, R.E. (2019) Evaluating the welfare of extensively managed sheep. *PLoS One, 14*(6), e0218603. https://doi.org/10.1371/journal.pone.0218603

Oktay, M.M., Boğan, M., Sabak, M., Gümüşboğa, H., Hakmeh, W., & Eren, Ş.H. (2021). Traumatic injuries from sheep sacrifice during the Eid Al-Adha holiday: A prospective multicentered study. *European Journal of Therapeutics*, 27(1), 8–13. https://eurjther.com/index.php/home/article/view/205/123

Opdam, H.I., Federico, P., Jackson, G.D., Buchanan, J., Abbott, D.F., Fabinyi, G.C.A., Syngeniotis, A., Vosmansky, M., Archer, J.S., Wellard, R.M., & Bellomo, R. (2002). A sheep model for the study of focal epilepsy with concurrent intracranial EEG and functional MRI. *Epilepsia*, 43, 779–787. https://doi.org/10.1046/j.1528-1157.2002.04202.x

Peirce, J.W., Leigh, A.E., da Costa, A.P., & Kendrick, K.M. (2001). Human face recognition in sheep: Lack of configurational coding and right hemisphere advantage. *Behav Processes*, 55(1), 13–26. https://doi.org/10.1016/S0376-6357(01)00158-9

Peruffo, A. & Cozzi, B. (2014). Bovine brain: An in vitro translational model in developmental neuroscience and neurodegenerative research. *Frontiers in Pediatrics*, 2, 74. https://doi.org/10.3389/fped.2014.00074. PMID: 25072040; PMCID: PMC4090595.

RSPCA (2025). *Sheep mulesing.* https://www.rspca.org.au/key-issues/sheep-mulesing/

Silva, S.R., Sacarrão-Birrento, L., Almeida, M., Ribeiro, D.M., Guedes, C., González Montaña, J.R., Pereira, A.F., Zaralis, K., Geraldo, A., Tzamaloukas, O., & Cabrera, M.G. (2022). Extensive sheep and goat production: The role of novel technologies towards sustainability and animal welfare. *Animals*, 12, 885. https://doi.org/10.3390/ani12070885

Simões, J., Abecia, J.A., Cannas, A., Delgadillo, J.A., Lacasta, D., Voigt, K., & Chemineau, P. (2021). Review: Managing sheep and goats for sustainable high yield production. *Animal*, 15(Supplement 1). https://doi.org/10.1016/j.animal.2021.100293

Steagall, P.V., Bustamante, H., Johnson, C.B., & Turner, P.V. (2021). Pain management in farm animals: Focus on cattle, sheep and pigs. *Animals*, 11, 1483. https://doi.org/10.3390/ani11061483

Taylor, S. (2020). Animal crips. In S. Jenkins, K. Struthers Montford, & C. Taylor, C. (Eds.), Disability and animality (1st ed.) (pp. 13–34). Routledge. https://doi.org/10.4324/9781003014270-3

Time Out (2025). *The story behind the NGV's 'sad sheep painting' and its mysterious artist.* https://www.timeout.com/melbourne/art/the-story-behind-the-ngvs-sad-sheep-painting-and-the-mysterious-artist-who-painted-it#google_vignette

Trovatelli, M., Spediacci, C., Castellano, A., Bernardini, A., Dini, D., Malfassi, L., Pieri, V., Falini, A., Ravasio, G., Riva, M., Bello, L., Brizzola, S., & Zani, D.D. (2022). Morphometric study of the ventricular indexes in healthy ovine BRAIN using MRI. *BMC Veterinary Research*, 18(1), 97. https://doi.org/10.1186/s12917-022-03180-0. PMID: 35277171; PMCID: PMC8915498.

Worldostats (2025). *Sheep Population by country 2025.* https://worldostats.com/sheep-population-by-country-2025/

Yurtman, E., Özer, O., Yüncü, E., Dağtaş, N.D., Koptekin, D., Çakan, Y.G., Özkan, M., Akbaba, A., Kaptan, D., Atağ, G., Vural, K.B., Gündem, C.Y., Martin, L., Kılınç, G.M., Ghalichi, A., Açan, S.C., Yaka, R., Sağlıcan, E., Lagerholm, V.K. ... Özer, F. (2021). Archaeogenetic analysis of Neolithic sheep from Anatolia suggests a complex demographic history since domestication. *Communications Biology*, 4(1), 1279. https://doi.org/10.1038/s42003-021-02794-8. PMID: 34773064; PMCID: PMC8589978.

6

Goats: Mischievous and fun-loving characters

INTRODUCTION

Physical mishandling of animals at slaughter has received considerable attention, but emotional mishandling tends to be overlooked (Kumar et al., 2023). The slaughterhouse environment, including vocalisations and the odour of blood and waste, can elicit severe emotional distress triggered by olfactory, visual and audible stimuli. This is particularly evident among young animals born to a mother at or on her way to the slaughterhouse.

Once heard, never forgotten—the desperate bleating of a mother goat moments before she is slaughtered, and the confused bleating of her baby, born at the slaughterhouse and destined to die within hours. In Australia, much of the goat meat is derived from free-living bush goats who roam the land until captured and transported by trucks for long distances to slaughterhouses. Often a pregnant goat will give birth on the journey or at the slaughterhouse under traumatic circumstances. Wild goats seclude from their herd immediately before and after giving birth, and newly born kids remain hidden for the first few days (Spitzer et al., 2022). Imagine the terror these sentient animals must feel when giving birth on the packed transport truck or in the crowded confines of the holding pen prior to entering the kill room. The newborn kids are collected and dumped into a small pen, where they take their first steps on wobbly legs while crying for the mothers who are already dead. Before long, the newborns also lie dead or left struggling for hours after a handheld bolt gun shoots a retracting bolt into their brain. This was Techno the kid's destiny until the team from J&D's Farm Sanctuary in Australia intervened.

Techno, a new beginning—J&D's Farm Sanctuary, Australia

Techno was born at a slaughterhouse and destined to be one of the nameless kids killed by a bolt gun to his fragile skull. Instead, he would become the first of around 70 newborn kids rescued from slaughterhouses by J&D's Farm Sanctuary in Northern Victoria, Australia.

DOI: 10.1201/9781003518853-7

Like many farmed animal sanctuaries, the motivation to found J&D's sanctuary lay in the love of one particular animal, Gussy girl the goose (J&D's Farm Sanctuary, 2025). The acquisition of 10 hectares (25 acres)—with eight trees and a lot of broken fences—was only the beginning of a journey to transform this land into an oasis for rescued animals. Amid tree planting and shelter building, the abused, neglected or abandoned farmed animals started arriving, reaching around 200 within a few short years. Gussy girl did not live to see the completion of the sanctuary she inspired, but her legacy lives on in the humans who changed their lives to provide safe haven for animals escaping intensive animal agriculture. Techno's story is shared by Deb who along with John founded J&D's Farm Sanctuary in 2021.

Techno's new life at the sanctuary got off to a rocky start as he patiently had to endure the discomfort and pain of daily treatments for some puzzling wounds on his lower back legs. Dosed up with antibiotics and pain relief, Techno's wounds needed flushing and sterilising, then dressed with "jelonet" paraffin gauze dressing and vet wrap every day until they healed. How these wounds were inflicted on an infant within hours of birth remained a mystery, but it seemed likely that he had been dropped into one of the holding pens with a grated floor. As he struggled to take his first steps, his tiny back legs may have slipped through the grates, and his skin and tendons shredded as he struggled to extricate himself.

It did not take long for the tenacious black and white kid to wriggle his way into everyone's hearts as he embraced the new life he had fallen into. Techno's unique personality and capacity to love showed none of the trauma inflicted on him by humans at such an early age. Research consistently reports that goats have excellent long-term memories, advanced cognition and emotional capacity if provided the opportunity (Harwood, 2019). Whether Techno retains memories of his early days will never be known. Possibly the sights, sounds and smells of a slaughterhouse may trigger these memories, but that would never happen once he found his forever home at J&D's Farm Sanctuary with goat friends Augustus and Agatha.

Techno thrives on interspecies sociability, finding joy with other goats in Goat Town, the sheep in the laneway and the humans who saved his life. As an infant, he would head butt everyone out of the way then curl up in his human's arms, audibly sigh with relief and soak up the comfort. He loved to be held all the time, seeming to misunderstand that as he grew in size, the human's arms did not. Now fully grown and living in Goat Town, Techno retains his insatiable capacity to find joy in every day as he rushes to the side of his humans whenever they are working in the paddocks.

Sadly, Techno's background story is not unique, but he did provide the impetus to continue saving newborn goats from the slaughterhouse. While not all survived due to hypothermia, infection, lack of feed—or stress and trauma—those who did went on to display their own unique personality and capacity to love and be loved as the irrepressible Techno.

Figure 6.1 Techno finds joy in every day and always needs to know what is going on. (Image supplied by J&D's Farm Sanctuary and reproduced with their permission.)

SOME BACKGROUND HISTORY

The goat's versatility saw it become one of the earliest domesticated animals, starting over 10,000 years ago and resulting in more than one billion goats inhabiting all types of ecological areas around the world (FAO, 2025). Their ability to adapt quickly to adverse conditions also meant the goat was taken for granted by humans during this period (Stilwell, 2017). Goats have undergone one of the largest increases in population growth since 2000, more than double that of pigs, cattle and sheep (Bertolini et al., 2018). In rural areas such as Asia and Africa where over 90 per cent of goats are distributed, they have been labelled the poor person's cow, being used for milk and cheese production, meat, manure, skin and fibre commodities, leather products, mohair wool, cashmere hair, transport as well as playing a sacrificial role in religious and cultural ceremonies (Amills et al., 2017). Goats were often traded in informal markets and slaughtered using unregulated traditional methods with no regard for animal sentience or welfare.

How the mischievous, fun-loving goat transformed into a provider of so many products for human consumption and a key player in the history and culture of human civilisation is a long story with parts still open to debate. Domestication is a cumulative and gradual evolutionary process resulting in the establishment of a new phenotype suitable for captivity and human needs. The stocky, gregarious goat antelope, the Bezoar goat, is widely believed to be the wild ancestor of the modern domestic goat (*Capra hircus*) with some archaeological evidence suggesting that domestication started in the Fertile Cresent area of the Middle East. More recent evidence suggests that there was a secondary site in Pakistan for the Cashmere-like goat breeds (Peng et al., 2022).

Following domestication, goats underwent a population expansion, especially in Asia and Africa, where the vast majority of the world's one billion goats still reside in a range of environments (Amills et al., 2017). They can be found in humid, tropical forests, cold, low-oxygen high-altitude regions, dry deserts and isolated small islands. This ecological diversity necessitated intense natural selection for adaptive traits, making goats particularly valuable to smallholders in areas that required hardiness and adaptability to harsh environments. The top five goat producers globally are India, China, Nigeria, Pakistan and Bangladesh, with smaller populations residing in Europe, America and Oceania (FAO, 2025). A second population expansion of some productive breeds, mostly of European origin, occurred with the technological developments of the nineteenth and twentieth centuries, resulting in the second phase of global gene flow (Lenstra, 2015).

Despite being one of the oldest domesticated animals, for many years, the goat avoided the industrialised agricultural practices that isolated farmed animals from their seminatural or natural free-range environments (Nomura et al., 2013). Goats prefer to live in large complex social systems where they form important dominance relationships, alliances and demonstrate reconciliation behaviours after fights (Briefer & McElligott, 2013). Segregating them in intensive farm settings raised significant welfare risks that necessitated increased selective breeding for desired traits.

Domestication of goats eventually resulted in the differentiation of goat breeds into two main types: selection based on different breeding systems and purposes, and selection based on different environments. Purposes included specific production traits such as milk production, meat or fibre, while environmental adaptations allowed goats to survive in a broad continuum of temperatures, rainfall, vegetation and water availability (Bertolini et al., 2018). The goat's survival and productivity in harsh climates, combined with their adaptability to novel situations and ability to cope with some stressors, made them significant contributors to the socio-economic development of rural economies where natural selection for traits compatible with the environment rather than selective breeding predominated.

Less than two per cent of the world's goat population lives in Europe, and yet the development of highly selective dairy breeds such as the Saanen, Alpine and Toggenburg under semi-intensive or intensive conditions resulted in around 17 per cent of the global goat milk and over 40 per cent of the global cheese market (Amills et al., 2017). In France, selection of traits for improved milk production focused on aspects of protein and fat content, while Spain established programmes of genetic improvement for dairy breeds to boost milk production per lactation.

GOATS IN CONTEMPORARY SOCIETY

The establishment of intensive dairy production systems in Europe and the northern countries of the Mediterranean prompted the European Commission's Food Safety Authority (2021) to investigate their welfare at slaughter. Goats had not received the attention of other more commonly mass slaughtered animals, instead being subject to a range of different and often inhumane methods. The resultant 2021 report recommended that goats should not be kept in lairage (holding pens) for a period prior to slaughter and should be provided basic welfare considerations such as access to water if holding was unavoidable. The report stipulated what should have been commonsense cues to help determine if goats were experiencing pain and fear during the lead-up to and slaughter process, thus highlighting how little goat welfare had featured in the past. These cues included slipping, falling, escape attempts, vocalisations, injuries, reluctance to move and turning back. Painful handling such as lifting, dragging by the horns or one leg, hitting with a stick and the use of dogs were singled out to be avoided, a bitter reminder of the potential pain and suffering these sentient commodified animals regularly received to meet human consumption requirements.

Goats, along with sheep and deer, are collectively described as small ruminants and often raised in semi-intensive farming systems. However, consumer perceptions of greater welfare with the inclusion of aspects of traditional grazing and breeding do not automatically eventuate (Tiezzi et al., 2019). Access to pasture allows more freedom of movement and expression of normal social behaviour, both of which are fundamental to meet Domain 4 of the Five Domains of animal welfare, but the potential for positive experiences is also required for overall welfare. With much of the contemporary research into goat welfare focused on dairy goats kept in intensive production systems and the development of welfare assessment protocols, less time has been spent assessing the welfare of bucks, kids, meat or fibre goats kept for long periods in semi-intensive systems (Minnig et al., 2021), and the wild free-ranging goats rounded up for slaughter.

What the research does suggest is that as intensive farming systems expand, goat welfare may be following a similar negative pathway as the dairy cow process. The goat may be hardy in natural settings, but confinement can set off a chain of negative outcomes, including vulnerability to infectious diseases such as Caprine Arthritis Encephalitis Virus (CAEV), lower quality of milk, poor overall health, increased environmental pollution from confined animals and aggressive behaviour (Tiezzi et al., 2019). Even if space allowance and provision of food and water per animal meet relevant standards, there is evidence that close proximity and absence of pasture may still result in welfare concerns among these social, nomadic animals.

Whether to immediately remove newborns, as it tends to happen in the dairy cow industry, has merited some research related to quality of milk rather than the attachment bond and wellbeing of mother and kid. With evidence that the mother–young bond has a strong effect on the regulation of hormones involved in lactation and that interruption of this bond immediately after birth can result in negative effects on lactation persistency and less effective oxytocin release during milking, there are economic grounds to keep mother and young together (Högberg et al., 2021). The complete emptying of the udder relies on oxytocin-mediated milk ejections, which then stimulates further milk secretion. Confirmation that oxytocin levels in plasma increased more during nursing and that higher milk yields resulted from suckling than machine milking presents a strong case to maintain this important bond, although some producers still choose to separate mother and kid.

Goats are renowned for their magnificent horns, but they are also perceived as a problem under intensive farming conditions. Kids, especially females from milking herds, are routinely subjected to disbudding to avoid potential injury to other goats, or to themselves from entanglement in equipment in the milking shed, or to humans when horns become too long (Singh et al., 2024). This usually occurs during their first week of life, before the horn buds have fused to the underlying frontal bone. Several methods have been used in the past, including thermal cautery through the generation of heat in a metal wire electrode to cut tissue. For anyone who has ever touched a hot iron or oven tray, it seems obvious that this painful process would leave a lingering deep pain once the heated metal electrodes were removed. There was also a risk of thermal brain damage and death from this significant surgical process if not conducted by experts such as veterinarians and with pain relief routinely provided. Unfortunately, the smaller global domestic goat industry compared to cattle has not only seen less consistent regulations, but also fewer drugs registered specifically for goats and fewer veterinarians specialising in goat medicine and surgery (Porter, 2024).

What is clear is that these sentient animals are capable of feeling pain during disbudding. Graphic images provided by animal activist organisations such as Farm Transparency Project (2019) showing young kids screaming and struggling as their budding horns are burnt off leaving the viewer in no doubt of this fact. Pain and distress stimulate the release of nociceptive neuroendocrine transmitters—just like humans—with a negative flow-on effect to the welfare and quality of life of the kids subjected to this procedure.

Attempts to selectively breed for polledness (absence of horns) as has happened with cows have not worked well for farmers. Dairy goats of European lineage, including the Saanen, Alpine and Toggenburg breeds, selectively bred for the recessive polledness gene can have reproductive complications including infertility (Singh et al., 2024). Females with the polled genotype grow into infertile intersex goats, while males experience an increased likelihood of developing sperm granulomas (inflammation).

Mohair and cashmere

Consumer demand for goat milk is not the only way that goats have been commodified. Angora goats are prized for their long curly hair from which mohair, a luxury fibre found in high-end fashion products, is produced (Four Paws, 2025a). As prey animals, goats are terrified of being constrained, vulnerable and defenceless during the painful and stressful shearing process carried out by often time-constrained workers. South Africa, where mohair production dates back to the mid-nineteenth century, produces around half of the world's mohair, making it a significant contributor to South Africa's economy. Images from a People for the Ethical Treatment of Animals (PETA, 2025) Asia investigation raised serious welfare concerns about workers lifting goats off the floor by the tail and potentially breaking their spine, sitting on struggling goats, hauling them around by their legs and throwing them across the floor after shearing. Initiatives such as the Responsible Mohair Standard were implemented to provide guidelines for ethical mohair production (Farmers Magazine, 2023).

Cashmere goats, the majority of whom live in China, Mongolia and Tibet, are similarly prized for their ultra-fine, soft undercoat used in high-end fashion clothing and knitwear. This process requires the goat's legs to be tied together before the undercoat is removed with a sharp comb, which can cause painful infected injuries if the skin is pierced by harried workers (Four Paws, 2025b). Cashmere goats—and all goats in general—are curious and social animals. They can memorise and recognise visual shapes, demonstrate perspective taking, use indirect information and human pointing, follow gaze direction and have a very good long-term memory (Briefer & McElligott 2013). That means they can remember their tormentors' faces and the emotional trauma associated with their treatment, thus increasing the cumulative burden on their welfare with repeated exposure to these events.

As if confining, disbudding and shearing were not enough suffering inflicted on goats, they also have a long history of sacrifice as part of "essential religious practices" or cultural traditions. Sociologists Henri Hubert and Marcel Mauss (1981) in their 1898 seminal book *Sacrifice: Its Nature and Functions* described how the ceremony and sacrificial ritual, no matter how painful and distressing for the animal, was deemed crucial in establishing a means of communication between the sacred and the profane human world. Pressure to ban animal slaughter, especially the prolonged suffering of slitting the goat's throat, has gained momentum in some regions of the world, but must still overcome centuries of tradition and speciesism. In 2006, the Calcutta High Court ordered that the slaughter process be visually concealed for cleanliness in public spaces, arguing that the blood and offal provided an unsettling visual experience for visitors at the pilgrimage and tourist site in the city (Berti & Good, 2024). Placing the protection of human sensibilities above the goat's experience of being ritualistically slaughtered again devalues this fun-loving animal for human benefit.

The resilience and adaptability of goats, which turned them into the poor person's cow in many developing nations, have also contributed to their ability to thrive when deliberately released or escaping captivity. Whether let free by sailors in countries such as the Pacific Islands, New Zealand and Australia, or abandoned in parts of Britain and France when local breeds were superseded by more productive breeds, these goats quickly settled in and multiplied in many different habitats. They soon achieved pest status, subject to indiscriminate culls. Helicopter-based shooting (or aerial culling) of goats is common in New Zealand and Australia, especially in open environments rather than forested and rugged terrain (Cox et al., 2023). In Australia, feral goats have become a source of revenue for drought-stricken farmers and helped Australia become the world's largest goat meat exporters in 2023 (Meat and Livestock Australia, 2024). However, there are welfare implications in the way these bush goats, including pregnant females, are rounded up, transported and slaughtered, and the treatment of newborns such as Techno.

Whether feral or domesticated, a source of meat or milk, mohair or cashmere, goats are unique sentient beings with emotions and personalities that increasingly have been subjugated to human commodification.

FEELINGS AND EMOTIONS

As with other farmed animals, goats have neural and behavioural abilities associated with advanced cognitive and emotional capacity—when they are given the opportunity to utilise them. They can manipulate objects, have long-term memory, seek environmental challenges and can infer the location of a reward through a process of elimination (Harwood, 2019). They also love to play, often outsmarting the humans with whom they interact on farmed animal sanctuaries.

Play is a luxury that many farmed animals will never live long enough to enjoy (Papageorgiou & Simitzis, 2022). Neurological studies suggest that play is rewarding in itself, and that animals are motivated to play because of the pleasure that they experience, both short term and long term. There is an immediate impact on the playful animal's affect and a delayed benefit to survival when the behaviour helps the animal to develop and enhance somatic skills to counter future stressful events. Play behaviour is also emotionally contagious both within and between species and is therefore considered an important positive welfare indicator.

Playful behaviour on farmed animal sanctuaries is a daily occurrence once traumatised animals settle in, as the amazing rescue story of Gianna the goat, her relocation to Ippoasi farm sanctuary and subsequent birth of Belzebee and Black Philippa reveal. Belzebee and Black Philippa never knew the trauma experienced by their mother as they did what kids do so well—kicked up their heels with the sheer joy of being alive, never seeming to tire of exploring and testing the safe environment their mother had fought so hard to reach. The stories of Gianna and her kids, as well as the older goats Tiberio and Cisco, and their new lives at Ippoasi farm sanctuary in Italy are narrated later in this chapter.

Vocalisations and facial expressions

Just like so many other farmed animals, vocalisations can provide a wealth of information about a goat's emotional state. In the wild, goats live in complex societies where behavioural and vocal expression of emotions provides a means of regulating social interactions within the group (Briefer et al., 2015). The contact call, used to maintain contact at close distances, is the most common goat vocalisation and can be separated into closed-mouth bleats and open-mouth bleats packed full of information about individuality, age, sex, body size, kinship and even group membership. A group member vocalising their food frustration can elicit a head-orientation bias to the right side, while positive and negative emotional-linked vocalisations can affect the behaviour and cardiac responses in other goats (Kumar et al., 2023). Open-mouth bleats have been observed during food anticipation, but also when a goat is in pain or other negative situations, and both open and closed mouth bleats are lower in extent and range for positive emotions compared to negative emotions such as isolation from the group (Laurijs et al., 2021). Vocalisations allow mothers and kids to recognise each other long after birth, and mother goats can also remember their kids' calls for up to a year after separation, demonstrating the long-term memory of these charismatic animals. However, vocalisations are also indicators of stress and social isolation, with a high number of goats observed vocalising on intensive farm systems, suggesting the existence of negative arousal (Tiezzi et al., 2019).

Domestication has heightened the goat's sensitivity to human facial expressions of emotion, with one study reporting a lateralised response towards positive human facial expressions in keeping with the Valence Hypothesis of mammalian brains. Goats in this study preferred to interact with the positive facial expression images when they were on the right side of the test area, suggesting a differential engagement of the left-brain hemisphere for approaching happy human faces, and conversely, an engagement of the right-brain hemisphere for avoiding the negative images on the left side (Nawroth et al., 2018).

The Valence Hypothesis of mammalian brains suggests a left/right specialisation for positive/negative emotions, respectively, as opposed to the Right Hemisphere Hypothesis, which suggests that the right hemisphere is specialised for processing all emotions (Nawroth et al., 2018). As is often the case, some studies suggest that both hypotheses are not mutually exclusive, instead co-existing during affective perception (Wyczesany et al., 2018). Irrespective of which hypothesis is preferred, the important observation is that farmed animals such as goats are able to recognise and distinguish between happy and angry images of the same person, supporting the notion that domestication has enhanced the interspecies communication and emotion perception in animals other than companion animals where it is taken for granted. That means domesticated non-companion animals who are commodified and eaten have the same brain structures and abilities as the companion animals that people choose to take into their homes, potentially creating cognitive dissonance if a person stops to think about this paradox.

Bellegarde and colleagues (2017) took a different approach to facial recognition, seeking to determine if goats would show differences in their reactions to images of goats' faces taken in positive and negative situations. They identified that goats displayed more ears forward at images of negative faces, a behaviour previously observed in both negative valence situations among goats, such as when an animal was socially isolated, and high arousal situations. Acknowledging that goats can identify conspecifics in a negative situation and therefore a potentially negative state—and in fact pay more attention to them—has welfare implications for goats observing and awaiting slaughter. Kumar and colleagues (2023) identified that exposure of goats to the slaughter of conspecifics altered the emotional state of those goats, causing significant changes in neurobiological activity (recorded as changes in the EEG spectrum) and increases in blood glucose levels.

Acute emotional arousal at point of slaughter is inevitable as goats are confronted with a barrage of olfactory, visual and audible stimuli triggering an automatic matching state between individuals (Kumar et al., 2023). In natural settings, emotional contagion allows animals to regulate social interactions and rapidly exchange information between group members. For goats, fear would normally relate to the presence of a predator or natural disaster, thus reinforcing group cohesion through the regulation of shared emotional responses. Humans have introduced the slaughter processes as an additional stressor during which the goat undergoes severe emotional distress at the unfamiliar environment, the odour of blood and waste, including the semiochemicals in the blood of slaughtered animals (that is, chemical substances released by an organism that affects the behaviour of other individuals), crowded conditions and fearful vocalisations, all triggering emotional contagion with no adaptive value and no escape for the captive goat.

During their life prior to slaughter, there is mounting evidence that deprivations of space, resources, physical contact and natural activities can have a strong negative impact on a captive goat's emotional wellbeing. When these stressors become long-term and chronic, they can also elicit impairment of cognitive functions and risk of mood disorders such as anxiety and depression—no different to that experienced by humans trapped in long-term stressful situations (Briefer & McElligott, 2013).

While some physical outcomes may be difficult to reverse, there remains hope for some goats' mental health within the security of a farmed animal sanctuary. A study conducted at Buttercups Sanctuary for Goats in Maidstone, United Kingdom, identified that after two to six years of good care and security in the sanctuary, goats rescued from poor welfare backgrounds showed either positive moods (the females) or moods similar to goats who had never experienced poor welfare backgrounds (the males) (Briefer & McElligott, 2013). An interesting observation was that rescued female goats showed more optimistic biases than males, suggesting females and males may react differently in certain situations. Importantly, it raised hope that goats can recover from previously poor conditions, a possibility evident in the stories of three goats from Ippoasi animal sanctuary in Italy.

New beginnings, Tiberio, Cisco and Gianna—Ippoasi animal sanctuary, Italy

Safely tucked away in the Migliarino San Rossore Massaciuccoli Park, near Pisa, Italy, lies a sanctuary for abused, exploited and traumatised animals. Escaping almost certain death in a society that refuses to acknowledge their right to live as unique individuals, these refugee animals can begin a new life at Ippoasi animal sanctuary.

Ippoasi is a voluntary association where animals of all species co-exist in a social space of mutual solidarity, away from exploitation and the need to forfeit their bodies or lives to satisfy human demand (Ippoasi, 2025). Resident animals are free to wander and intermingle within the safety of Ippoasi, but to venture beyond this safe zone risks returning to the exploitation and death from which they escaped. For the first time in their lives, they can live in peace, respected as sentient beings with agency rather than dictated by the tenets of speciesism and the harmful belief that humans are the superior species.

In his 1977 book *Animal Liberation*, Australian philosopher Peter Singer who popularised the term speciesism argued against the perception that being human was sufficient to maintain privileges over animals, adding in a 2023 interview "….I don't think the species of a being is a reason for counting more than another being. What is important is the capacity to suffer and to enjoy life" (Corbyn, 2023).

As an anti-species sanctuary, Ippoasi rejects the traditional hierarchical vision of species and awards all animals equal moral standing. Tiberio, Cisco and Gianno are three of those animals whose lives changed dramatically when entering the safe, non-discriminatory world of Ippoasi. Susanna kindly shared the story of these three goats, each with a different background and personality, but bound by an unstoppable will to live when given that opportunity at Ippoasi.

Tiberio

Tiberio was an aggressive force to be reckoned with when he first came to the attention of Susanna and the Ippoasi team. And who could blame this magnificent brown billy goat? For ten long years, he was captive in the former Cavriglia zoo in Arezzo Italy, deprived of his freedom and self-determination. His frustration turned him into one of the most feared animals in the zoo. During the two years it took to secure his freedom, Susanna and the team developed a heathy respect for his huge horns and imposing physique.

When he first arrived at the sanctuary, Tiberio was furious with life and everyone who came near him. He checked all the fences and gates that stood in his way and such was his strength, all the gates needed reinforcing to prevent their destruction.

But what a difference only a few months made. His desperate rage-filled head butts turned into a gentle rub of his horns on a person's hips as way of greeting, and this gentle giant became everyone's friend within the safety of Ippoasi. Even when racing to eat up the vegetables left over from the pig feed, Tiberio was ever mindful of the power of his gigantic horns and ensured he hurt no one. But looking into Tiberio's eyes, it was still possible to see glimpses of a harrowing story of suffering, rage and loneliness buried beneath his resilience, and newfound joy and trust.

Tiberio lived at the sanctuary for seven years, becoming one of the most prominent personalities as he learned to express himself away from the cruel captivity he had endured for human entertainment. The majestic Tiberio, serene and without a care in the world as he "sang" to himself and his friends Capret, Tina, Rachelle and Ursula, was a constant reminder that the most beautiful things in life are free.

In 2020, Tiberio fell asleep and peacefully left this world as a free spirit. His story of resistance and liberation touched so many hearts, and his death caused grief among all those—animals and humans alike—who had the privilege of sharing Tiberio's transformative journey to freedom in a sanctuary where he was respected as a unique individual.

Figure 6.2 Tiberio became everyone's friend once he felt safe at Ippoasi. (Image supplied by Ippoasi animal sanctuary and reproduced with their permission.)

Cisco

Snow white Cisco was an imposing male breeder from a supposed ethical free-range farm that subsequently turned out to be an abusive environment no different to intensive farming systems. The commanding billy goat with huge horns had been held captive in a closed container where he was routinely beaten for lack of cooperation and aggression towards his captors.

When the farm went out of business, Cisco came to the sanctuary. He initially appeared resentful and intimidating due to his size, but how quickly things changed. Through the net of his separated enclosure as he awaited the desexing surgery that would ensure he was never again exploited for his body, Cisco was introduced to the wonder of goats, pigs, horses and other animals roaming freely around the sanctuary.

And with that Cisco transformed into one of the friendliest goats imaginable. From a grumpy billy-goat, turning his back on all who approached, Cisco became the first to approach the team, wandering over to say hello by rubbing his head on their legs. Of course, this slow meander would turn into a run if there were interesting buckets in the hands of the team.

Cisco also became the self-appointed animal ambassador and tour guide for visitors learning about the animal residents and the sanctuary's work. Escorting them meant he had a captive audience from whom he could demand attention and adoration, and with his zest for life, Cisco was irresistible and received both from all who met him. Visitors would leave the sanctuary with an indelible memory of the goat with the determined and proud pace, controlling the group of visitors with a sniff of the air and a toss of his majestic horns. The human tour guides soon learned to stifle their impatience at Cisco's stubbornness and determination to control the group, well aware that Cisco was not going to leave and was far more captivating than them.

From a breeding machine to create more goats destined for use, abuse and slaughter, Cisco became a true individual—curious, dynamic, bold—and friend to companions Capret, Tina, Palle, Amore and Angela, and with an abiding love for red onions.

Gianna

Early one morning, the sanctuary received a phone call from a military barracks a few minutes' drive from Ippoasi. This phone call would change the life of Gianna, a gentle goat desperate to find safety from whatever horrific events preceded her arrival in the army barracks dormitory. Terrified, exhausted and dizzy, Gianna had the remnants of a chewed rope compressing her neck, stark testimony to the desperation that caused her to flee her previous life.

A few hours later, the veterinary check confirmed what Gianna's shape had hinted at: Gianna was pregnant and, by good fortune, had stumbled into a safe haven where she could deliver her babies in peace. Although scared and trembling while these checks and life-saving therapies were undertaken, Gianna's bid for freedom would pay off when she gave birth to two kids who would never know the heartache of

Figure 6.3 Cisco is the self-appointed animal ambassador and tour guide for visitors. (Image supplied by Ippoasi animal sanctuary and reproduced with their permission.)

being torn from their mother's side. For probably the first time in her life, Gianna was able to become the protective and caring mother that is the right of all animals, rather than having her kids—and the milk she produced for them—stolen from her by humans. Gianna was never separated from her kids, remaining a tired but attentive mother to Belzebee and Black Philippa, the playful kids whose joy for life and confidence around humans were a constant reminder of how goats should live.

Gianna's full story will never be known nor the torments from which she escaped, but her desperate act of resistance and rebellion freed Gianna and her future infants from pain, separation and slaughter. This act of courage is duplicated worldwide among mothers who will do anything to protect their young—whether human or non-human—if given the opportunity.

Figure 6.4 Gianna with Belzebee and Black Philippa, the kids she was determined to protect. (Image supplied by Ippoasi animal sanctuary and reproduced with their permission.)

ETHICAL IMPLICATIONS

Since early domestication, the versatility of goats has seen them taken for granted by humans, culminating in some goats joining the ranks of other farmed animals in intensive commercial settings. While this has prompted a growing body of research, there are still many unknowns that can leave gaps in the welfare of goats in the many environments they now inhabit. To truly understand the goat in a range of contexts, humans need to pay attention to their behaviours, vocalisations, body language and social dynamics, all of which can convey feelings, needs and distress.

Like many farmed animals, goats are stereotyped, including beliefs that goats eat everything, only males have horns, goats are not very smart (reinforced by social media clips showing them doing weird and wonderful things), goats are low maintenance and more. Their name appears in the word "scapegoat", which describes someone who is blamed for the mistakes of other people, as well as the phrase "get your goat", which means to annoy someone. Somewhere along the way, the uniqueness of these cheeky sentient animals was lost to all but those who choose to really listen to them. On farmed animal sanctuaries, it becomes unavoidably evident that every goat is unique and ready to move on with living their best life if provided security and autonomy to be a goat and do goat things.

REFERENCES

Amills, M., Capote, J., & Tosser-Klopp, G. (2017). Goat domestication and breeding: A jigsaw of historical, biological and molecular data with missing pieces. *Animal Genetics, 48*(6), 631–644. https://doi.org/10.1111/age.12598

Bellegarde, L.G.A., Haskell, M.J., Duvaux-Ponter, C., Weiss, A., Boissy, A., & Erhard, H.W. (2917). Face-based perception of emotions in dairy goats. *Applied Animal Behaviour Science, 193*, 51–59. https://doi.org/10.1016/j.applanim.2017.03.014

Berti, D. & Good, A. (2024). The judicialisation and politicisation of sacrifice. In D. Berti & A. Good (Eds.), *Animal sacrifice, religion and law in South Asia* (pp. 3–52). Routledge.

Bertolini, F., Servin, B., Talenti, A., Rochat, E., Kim, E.S., Oget, C., Palhière, I., Crisà, A., Catillo, G., Steri, R., & Amills, M. (2018). Signatures of selection and environmental adaptation across the goat genome post-domestication. *Genetics Selection Evolution, 50*(57). https://doi.org/10.1186/s12711-018-0421-y

Briefer, E.F. & McElligott, A.G. (2013). Rescued goats at a sanctuary display positive mood after former neglect. *Applied Animal Behaviour Science, 146*(1–4), 45–55. https://doi.org/10.1016/j.applanim.2013.03.007

Briefer, E.F., Tettamanti, F., & McElligott, A.G. (2015). Emotions in goats: Mapping physiological, behavioural and vocal profiles. *Animal Behaviour, 99*, 131–143. https://doi.org/10.1016/j.anbehav.2014.11.002

Corbyn, Z. (2023, May 22). Philosopher Peter Singer: 'There's no reason to say humans have more worth or moral status than animals'. *The Guardian*. https://www.theguardian.com/world/2023/may/21/philosopher-peter-singer-theres-no-reason-to-say-humans-have-more-worth-or-moral-status-than-animals

Cox, T.E., Paine, D., O'Dwyer-Hall, E., Matthews, R., Blumson, T., Florance, B., Fielder, K., Tarran, M., Korcz, M., Wiebkin, A., & Hamnett, P.W. (2023). Thermal aerial culling for the control of vertebrate pest populations. *Scientific Reports, 13*, 10063. https://doi.org/10.1038/s41598-023-37210-0

EFSA AHAW Panel (EFSA Panel on Animal Health and Welfare), Nielsen, S.S., Alvarez, J., Bicout, D.J., Calistri, P., Canali, E., Drewe, J.A., Garin-Bastuji, B., Gonzales Rojas, J.L., Gortazar Schmidt, C., Herskin, M., Miranda Chueca,

M.A., Padalino, B., Pasquali, P., Roberts, H.C., Spoolder, H., Stahl, K., Velarde, A., Viltrop, A., Winckler, et al. (2021). Scientific opinion on the welfare of sheep and goats at slaughter. *EFSA Journal*, *19*(11), 6882. https://doi.org/10.2903/j.efsa.2021.6882

Farm Transparency Project (2019). *Goat disbudding*. https://www.farmtransparency.org/campaigns/goat-disbudding

Farmers Magazine (2023). *Goats and mohair: South Africa's role in the global mohair industry*. https://farmersmag.co.za/2023/05/goats-and-mohair-south-africas-role-in-the-global-mohair-industry/

Food and Agricultural Organisations of the United Nations (2025). *Crops and livestock products*. https://www.fao.org/faostat/en/#data/QCL

Four Paws (2025a). *Mohair*. https://www.four-paws.org.au/campaigns-topics/campaigns/wearitkind/mohair

Four Paws (2025b). *Cashmere*. https://www.four-paws.org.au/campaigns-topics/campaigns/wearitkind/cashmere

Harwood, D. (2019). Health and welfare priorities in goats. *Veterinary Record*, *185*(9), 264–266. https://doi.org/10.1136/vr.l5103

Högberg, M., Winblad von. W. L., Hydbring-Sandberg, E., Forkman, B., & Dahlborn, K. (2021). Growth rate and behaviour in separated, partially separated or non-separated kids and the corresponding milk production of their mothers. *Journal of Dairy Research*, *88*, 381–387. https://doi.org/ 10.1017/S0022029921000789

Hubert, H. & Mauss, M. (1981). *Sacrifice: Its nature and functions* (Revised ed.). University of Chicago Press.

Ippoasi (2025). *About us*. https://ippoasi.org/chi-siamo/

J & D's Farm Sanctuary (2025). *About us*. https://jdsfarmsanctuary.org.au/about-us/

Kumar, P., Abubakar, A.A., Ahmed, M.A., Hayat, M.N., Ajat, M., Kaka, U., Goh, Y.M., & Sazili, A.Q. (2023). Electroencephalogram and physiological responses as affected by slaughter empathy in goats. *Animals*, *13*, 1100. https://doi.org/10.3390/ani13061100

Laurijs, K.A., Briefer, E.F., Reimert, I., & Webb, L.E. (2021). Vocalisations in farm animals: A step towards positive welfare assessment. *Applied Animal Behaviour Science*, *236*, 105264. https://doi.org/10.1016/j.applanim.2021.105264

Lenstra, J. (2015). Origin and history of livestock diversity. In B. D. Scherf & D. Pilling (Eds.), *Second report on the state of the world's animal genetic resources for food and agriculture* (pp. 5–23). Food and Agriculture Organization of the United Nations. https://openknowledge.fao.org/server/api/core/bitstreams/464ecee6-d55c-407d-93bb-e9890ee88058/content

Meat and Livestock Australia (2024). *2024 Global goat snapshot released*. https://www.mla.com.au/news-and-events/industry-news/2024-global-goat-snapshot-released/

Minnig, A., Zufferey, R., Thomann, B, Zwygart, S., Keil, N., Schüpbach-Regula, G., Miserez, R., Stucki, D., & Zanolari, P. (2021). Animal-based indicators for on-farm welfare assessment in goats. *Animals*, *11*, 3138. https://doi.org/10.3390/ani11113138

Nawroth, C., Albuquerque, N., Savalli, C., Single, M-S. & McElligott, A.G. (2018). Goats prefer positive human motional facial expressions. *Royal Society Open Science*, *5*, 180491. https://doi.org/10.1098/rsos.180491

Nomura, K., Yonezawa, T., Mano, S., Kawakami, S., Shedlock, A.M., Hasegawa, M., & Amano, T. (2013). Domestication process of the goat revealed by an analysis of the nearly complete mitochondrial protein-encoding genes. *PLoS One*, *8*(8), e67775. https://doi.org/10.1371/journal.pone.0067775

Papageorgiou, M. & Simitzis, P.E. (2022). Positive welfare indicators in dairy animals. *Dairy*, *3*, 814–841. https://doi.org/10.3390/dairy3040056

Peng, W., Zhang, Y., Gao, L., Feng, C., Yang, Y., Li, B., Wu, L., Wu, A., Wang, S., Ren, X., Chen, Z., Zhang, M., Cai, D., Wang, X., Lv, M., Zhang, Y., Li, S., Zhang, Y., Huang, L. & Li, S. (2022). Analysis of world-scale mitochondrial DNA reveals the origin and migration route of East Asia goats. *Frontiers in Genetics*, *13*, 796979. https://doi.org/10.3389/fgene.2022.796979

PETA (2025). *Groundbreaking Exposé: Goats Thrown, Cut, and Killed for Mohair*. https://investigations.peta.org/goat-mohair-expose/

Porter, T. (2024, March 18). Disbudding goat kids. *Vet Practice*. https://www.vetpracticemag.com.au/disbudding-goat-kids/

Singer, P. (1977). *Animal liberation: A new ethics for the treatment of animals*. Avon.

Singh, P., Venkatachalam, D., Kongara, K., & Chambers, P. (2024). Pain mitigation strategies for disbudding in goat kids. *Animals*, *14*, 555. https://doi.org/10.3390/ani14040555

Spitzer, H.B., Meagher, R.K., & Proudfoot, K.L. (2022). The impact of providing hiding spaces to farmed animals: A scoping review. *PLoS One*, *17*(11), e0277665. https://doi.org/10.1371/journal.pone.0277665

Stilwell, G. (2017). Why and how to measure goats' welfare. In J. Simões & C. Gutiérrez (Eds.), *Sustainable goat production in adverse environments, Volume 1* (pp. 439–453). Springer. https://doi.org/10.1007/978-3-319-71855-2_25

Tiezzi, F., Tomassone, L., Mancin, G., Cornale, P., & Tarantola, M. (2019). The assessment of housing conditions, management, animal-based measure of dairy goats' welfare and its association with productive and reproductive traits. *Animals*, *9*(11), 893. https://doi.org/10.3390/ani9110893

Wyczesany, M., Capotosto, P., Zappasodi, F., & Prete, G. (2018). Hemispheric asymmetries and emotions: Evidence from effective connectivity, *Neuropsychologia*, *121*, 98–105, ISSN 0028-3932. https://doi.org/10.1016/j.neuropsychologia.2018.10.007

7

Chickens: Foraging with friends

INTRODUCTION

"The sky's a-falling" cried Henny-Penny after being hit on the head with an acorn (Jacobs, 1892). In this English folktale, the brave chicken Henny Penny rushes to warn the king of the imminent end of the world, along the way collecting a rooster, duck, goose and turkey, all of whom believe Henny-Penny's panicked prediction. Even the fox they meet appears to believe them, until the stereotypical superiority of mammals and stupidity of birds is validated when the fox leads them into a hole and all but Henny-Penny meet a grisly death.

Perpetuation of the foolish "bird-brain" stereotype, even from childhood, can serve a protective function against cognitive dissonance. It would be harder to confine or slaughter billions of chickens annually if consumers did not perceive them as a mindless, unfeeling source of food. This raises the troubling question: Would humans continue to commodify chickens if they recognised and accepted their cognitive and emotional abilities? Sadly, the answer is probably yes as what chickens offer to some people in the form of meat and eggs provides adequate justification to imprison and destroy these sentient beings.

Neuroscientist and animal advocate Lori Marino (2017) provided an insightful exploration of domestic chickens when commenting:

> Unlike many other birds, chickens are categorised as a commodity, devoid of authenticity as a real animal with an evolutionary history and phylogenetic context. Thus, arguably, perceptions of chickens shape their use as commodities which, in turn, then reinforces those original perceptions.
>
> *(p. 127)*

Spend some time with chickens, and it becomes apparent that they can show happiness, frustration, boredom, attachments and grief. They are sentient animals

DOI: 10.1201/9781003518853-8

with well-developed social skills and unique personalities when given the opportunity to be recognised as individuals.

Globally, domestic chickens outweigh all wild birds by a factor of three, and for every five wild birds, there is one chicken (Edison & Esvelt, 2020). That allows poultry in general to achieve the highest annual global per capita consumption of over 14 kilograms (31 pounds), higher than pork, beef and sheep (Siddiqui et al., 2023). Humans clearly love chickens and what their bodies offer.

There are the broiler chickens, those selectively bred for a high meat content, which may be reached at an unnaturally early period in their development. Approximately 70 billion broiler chickens are slaughtered annually, making broiler chicken farming the largest terrestrial animal production system in the world (Wilcox et al., 2024). Layer hens are different to broiler meat chickens. When a batch of eggs hatch from the breeder hens, the females are destined to move to the rearing facility, followed by the laying facility. Most males face immediate death, given their worthlessness as egg layers. Once the females reach a certain weight, they move to the laying facility, where increased light exposure, coupled with induced moult by diet and environmental manipulations, stimulates increased and relentless egg production for the entirety of their lives. With all these manipulations, it is not surprising that the chicken's natural ten-year lifespan is cut to around 72 weeks as commercial egg layers.

The relentless demand for chicken meat and eggs has resulted in some of the cruellest practices in intensive animal agriculture and the failure to recognise single chickens as unique individuals. While a healthy flock is crucial for economic success, the monetary value of an individual chicken can be inconsequential. In a shed containing thousands of chickens, casualties are an inevitable consequence—perhaps not so surprising in an industry where day-old male chicks are macerated in a high-speed grinder or asphyxiated by carbon dioxide simply because they do not lay eggs. The fate of male chicks is discussed in the next chapter.

While individual deaths may be inconsequential in the poultry industry, more attention is paid when deaths total in the hundreds of thousands, as can happen in extreme weather events or outbreaks of Avian Influenza. When Highly Pathogenic Avian Influenza (HPAI) H5N1 spread from aquatic birds to domestic poultry in 2022, domestic poultry producers globally faced economic ruin (CDC, 2024). In 2023 and 2024, gruesome images of millions of chicken carcasses being tipped into containers epitomised the miserable life and death of confined chickens whose living arrangements left them vulnerable to disease. Reporting their deaths in terms of financial loss and insurance claims for the producers, and egg shortages and increased prices for consumers, gave no recognition to the traumatic death of these unique individuals trapped in a cruel production system.

SOME BACKGROUND HISTORY

The exact origins of the domesticated chicken have been debated, but there is some evidence that over the last eight to ten thousand years, the suspected progenitor, the Red Junglefowl (*Gallus gallus*), was selectively bred to become the domestic chickens of contemporary times. Ironically, while the Red Junglefowl is now endangered in the field edges, groves, scrubland, forests and semi-open habitats of south-east Asia and the Indian subcontinent, the chicken is one of the most widespread domestic animals.

The first domesticated chicken remains were found in China and Pakistan around 7,500 years ago, spreading to Russia, Turkey, parts of eastern Europe, then Europe and North America. Wherever they went, their natural behaviours were enacted in highly social groups with strict hierarchies (Ericsson, 2022) until ancient Egypt developed huge incubators that could hatch up to 15,000 chicks simultaneously and maintain them at an optimal temperature to maximise production (Jones, 2014). The versatile chicken had begun its transition to a faceless commodity for use and abuse according to human need.

Domestication of animals requires selection for reduced fearfulness of humans, and chickens were no exception. The Red Junglefowl remained vigilant of humans and, as prey animals, maintained better spatial learning and higher imprinting abilities and behaviour flexibility among chicks compared to domestic breeds. However, some behaviours could not be bred out of domestic chickens, as anyone who has observed chickens wandering freely will attest. Their love of foraging, scratching and dust bathing is fundamental to their wellbeing, similar to Red Junglefowls, who spend the majority of their waking time scratching or pecking at the ground, preening, dust bathing and perching on branches, especially at night (Wilcox et al., 2024). The frustration imposed on chickens who have retained these basic motivations despite being confined in barren intensive farm settings is physically and emotionally detrimental.

British ethologist Marian Dawkins (1983) highlighted the urgent need for hens to follow their instincts for "dust bathing", a process where they work dust, sand or some sort of loose material into their feathers to absorb excess oil. Dawkins demonstrated that a chicken's motivation for dust bathing could be as strong as their motivation to eat when offered the choice of entering a cage with dust bathing material or food (Fraser & Weary, 2022). Where environmental restrictions prevented these natural behaviours, deflection into detrimental behaviours such as feather pecking and cannibalism became a real possibility. The Red Junglefowl also enjoyed life in small groups of mixed sexes, something domesticated chickens could not experience in confined indoor commercial flocks of tens of thousands of chickens (Garnham & Løvlie, 2018).

During their thousands of years of domestication, populations of the Red Junglefowl developed independently in different parts of Asia, encouraged by selective breeding to meet specific human demand. Cock fighting and religious slaughter fulfilled entertainment and cultural needs, but as industrialised animal farming took hold to meet global demand for proteins, selected traits were no longer conducive to the long-term health and wellbeing of chickens. Fast food outlets, such as KFC, Red Rooster, Oporto and McDonalds with their fried chicken, chicken nuggets, grilled chicken and coated chicken, ensured a relentless demand for broiler chickens designed to grow at extremely fast rates.

Domestication in a range of animals and birds has seen reduced brain size proportionate to body size compared to their wild relatives and forebears. This is believed to reflect the reduced functional needs of domesticated animals' brains when removed from their natural setting, but the unnatural body size augmentation in chickens can confound these ratio calculations (Henriksen et al., 2016). When researchers Henriksen and colleagues (2016) compared brain mass and body mass in domesticated White Leghorn chickens and Red Junglefowl, they found that the domesticated chickens' body mass had increased by approximately 85 per cent during domestication, and their brain mass had increased by approximately 15 per cent, mainly due to enlargement of the cerebellum. They proposed that reduced relative brain mass during domestication was caused by increased body mass, thus contesting the common perception that domestication had reduced the overall brain mass of domestic chickens. Perceiving domesticated chickens as less smart and with reduced self-awareness may alleviate some of the tensions of cognitive dissonance, but also raises the question that even if consumers chose to recognise their potential capabilities, would chicken welfare improve?

CHICKENS IN CONTEMPORARY SOCIETY

Imagine being trapped in a body that has grown so quickly your legs can no longer support your weight. Walking becomes increasingly difficult, forcing you to spend your life sitting in painful frustration until your body reaches the optimum size for slaughter. For many meat chicken breeds genetically selected for rapid growth, this is their reality as they will never experience perching, ground scratching, dust bathing and walking. Between 1957 and 1991, the growth rate of broiler chickens increased by up to 400 per cent (Havenstein et al. 2003), with around 90 per cent of this increase a direct result of selective breeding and the remainder from advancements in nutritional products. Rapid and increased pectoral mass to provide the massive breast meat chunks for supermarkets and fast-food outlets saw juvenile broiler chickens' immature bone structures, especially in the legs, twisting, buckling and crumbling under the unnatural weight. These malformed chickens' short lives were marred by chronic pain, cardiovascular problems from the inevitable strain on the heart and increased risk of Sudden Death Syndrome. In addition, aggression, stereotypies including pacing, pecking at empty feeders, over-drinking and feather pecking, as well as elevated corticosterone levels, were indicative of suffering and frustration during a broiler chicken's short life. Taking commodification to a new level, in some Asian countries such as Malaysia and Indonesia, farmers are contracted to "grow" chickens by vertical integrator companies, with harvesting after a month or so when birds reach two kilograms in weight (Wilcox et al., 2024).

Layer hens are also subject to multiple welfare and ethical issues when diet and environmental manipulations force them to maintain an unnatural schedule of producing more and more large eggs. Conventional cage rearing systems for laying hens emerged after the Second World War as a way to increase production. When the relationship between increased egg laying and increased light hours became evident, the fate of many layer hens was sealed for decades to come (Jones, 2014). While conventional battery cages that force layer hens to live in small barren wire cages remain legal in some regions, there is a global push to phase them out in favour of furnished cages and non-cage free range or organic systems (Bonnefous et al., 2022). However, this is proving a slow process with lengthy changeover timelines granted to some producers in recognition of the expenses entailed. Many generations of layer chickens remain doomed to live out their lives in confinement with automated feed and watering, ventilation, lighting, manure and egg collection systems. Foraging, scratching, dust bathing, nesting, perching and even stretching their wings remain impossible for these chickens churning out egg after egg until "spent". Spent chickens are worn out after around 18 months of laying up to 300 eggs a year, and once their egg production declines and they are no longer profitable to keep, they face slaughter and replacement with younger productive chickens.

Recognising the cruelty of slaughtering spent hens, charities in some countries have been established to rehome ex-commercial laying hens. For example, the British Hen Welfare Trust rehomes around 60,000 hens each year, with their one-millionth rescue hen rehomed to Highgrove Gardens, the property of King Charles III of England, in 2024 (Ng, 2024).

Consumer concerns and animal advocacy have seen an increase in alternative housing methods for chickens, especially in many western nations. Developed as an alternative to battery cages, furnished or enriched cages claim to offer more space and height, plus provisions for normal behaviours such as nesting, scratching and perching. However, according to animal advocates Animal Justice, some enriched cages appear to be tokenistic attempts to improve welfare when chickens remain in overcrowded conditions with high morbidity and mortality rates and insufficient space for vital natural behaviours such as dust bathing and foraging (DeMarsh, 2024).

In the barn system, hens can move around large sheds with different levels. Thousands to tens of thousands of birds may reside in each shed, and these densities determine if and how normal behaviours and social systems can be expressed. With high numbers of unconfined chickens comes the risk of them tearing at each other's feathers in an unstoppable struggle for dominance. Free range requires chickens to have access to an outdoor range at least during the day, allowing them to forage, walk around, stretch their wings and dust bathe. With access to both the barn and the range, the hens gain some decision-making capacity as to whether to explore the range or to stay in the barn (Ferreira et al., 2021). While the thought of happy chickens may alleviate consumer guilt and provide excellent marketing material, the chicken's life still culminates in use, abuse and death for human consumption.

In 2006, blogger Ben Grossblatt drew attention to the advertising images of happy animals who seemed delighted to be fattened up to die for human consumption. Describing these animals as "suicide food", Grossblatt (2011) pointed out the cruel irony of depicting cows, hens and other animals in peaceful rural settings and seeming to delight in preparing their bodies to be eaten. For urban dwellers with limited knowledge or contact with farmed animals, advertising campaigns of happy hens wandering around a farmyard or choosing where to lay eggs can allay the inevitable cognitive dissonance if forced to confront the truths behind the animal agriculture industry.

The commodification of these smart and social creatures ignores the growing evidence-base supporting their cognitive abilities, self-control, empathy and abilities to adapt to a social, complex and constantly changing environment (Ferreira et al., 2021). Those hatchlings permitted to live can recognise partially blocked objects within days of hatching, and even remember that an object exists when it is shifted out of sight. As chickens mature, their cognitive skills rival some mammals, including memory of personal experiences related to food as well as self-control when passing up an immediate reward in favour of a delayed greater reward—something that infant children take years to develop (Marino, 2017).

Backyard chickens

There is some evidence that red junglefowls may have started their relationship with humans as pets rather than a food source—especially when the newly hatched chick imprinted on a human as the first large object they observed (Coulthard, 2022). During the nineteenth century, exotic breeds of chickens became popular status symbols among the wealthy in some western countries. This trend slowed down in the twentieth century with the large-scale commercialisation of chickens for eggs and meat. By the twenty-first century, non-commercial chickens had again achieved popularity in western countries such as the USA with an estimated 11 million backyard chickens (Greene, 2025) and around 400,000 in Australia (Cosgrove, 2025). In 2021, chickens were the sixth most popular pet in the United Kingdom, with approximately 1.2 million chickens (Baldrey et al., 2024). Backyard chicken flocks also increased in countries across Europe, including France where reasons for their popularity included producing healthy and local products, educating children about the food they eat and seeking a pet bird (Souvestre et al., 2021).

Commercial chicken welfare has faced increased scrutiny and advocacy, but non-commercial chickens kept in a domestic backyard environment occupy a different welfare and legal space. Globally, chickens are the second most kept class of birds, behind the parrot family, with reasons ranging from food to company (RSPCA, 2025). They are sometimes perceived as a low-maintenance pet, despite them needing appropriate food, mental stimulation, security, healthcare and chicken companions. Scratching and foraging can destroy a beautifully manicured garden, and their welfare can suffer when people do not accept them as quirky and unique individuals, not just egg providers.

Having evolved as prey animals, the chickens' natural survival instinct is to hide signs of illness from potential predators—including humans—making it essential that those who keep chickens understand their body language. Researchers Singleton and colleagues (2021) examined the United Kingdom veterinary records and found euthanasia rates among chickens to be 30 times higher than dogs, 15 times higher than cats and seven times higher than rabbits. Some possible reasons included the chickens' stoicism making it too late to assist, fewer medical interventions being available for chickens or some chickens being perceived as more disposable compared to traditional dog and cat companion animals. Backyard chickens seemed to occupy an uncertain space with names such as "pets with benefits" and "productive backyard pets" identified in this study. A strange dichotomy was also identified among some backyard chicken owners who loved their own chickens and would never eat them but continued to eat chicken from shops or fast-food outlets.

The relatively low cost of purchasing chicks, coupled with ease of access ranging from backyard breeders, marketplaces, online sources and free chicks hatched in hatchling programmes in schools and other facilities, raises potential ethical concerns for their welfare in life and death, as the story of Zola shows.

Zola's story—Mother Clucker's Microsanctuary, USA

"Free hen so she can have a second chance at life by being someone's dinner" read the bizarre ad on the classified advertisement website. Raised by a backyard chicken keeper, the chicken who would be named Zola was the sole survivor of a predator attack on the flock, but her injuries were extensive. Zola did receive a second chance at life—but not on someone's dinner plate—when she was rescued by Anissa from Mother Clucker's Microsanctuary in North Carolina, USA (Mother Clucker's Microsanctuary, 2025). Putting Zola on a dinner plate was never an option once she reached the safety of the vegan sanctuary, where each chicken was loved and respected for their unique personality, not their body.

Zola arrived at Mother Clucker's in a cardboard box, her body wracked with pain and her eyes filled with fear and defeat. The attack had left her scalped, half her upper beak missing and several deep wounds. Her comb and head skin had already turned necrotic, but scaly leg mites and internal parasites made it evident that the abuse and neglect of this backyard chicken had started long before the predator attack.

Anissa reassured the bedraggled brown chicken that she mattered and was now safe as she gently cleaned her painful wounds while awaiting the avian veterinarian. Zola could not understand the words, but the tone—and the soft blankets beneath her battered body—seemed to allay some of the anxiety and fear in this sentient animal whose value had been measured by the eggs she produced.

Zola's future was grim, with a reproductive mass found in her body compounding the horrific wounds. The more eggs laying chickens produce, the more risk they face for reproductive disease from this unnatural burden on their bodies. Zola's body temporarily shut down her egg laying while healing, allowing her to benefit from the antibiotics, pain medications, liquid food necessitated by the damage to her beak and Anissa's constant verbal and tactile reassurance that she was safe and loved.

Chickens communicate using at least 24 distinct vocalisations (Marino, 2017), and before long, Zola was proving this as she would squeal with excited chatter every time Anissa approached. Despite whatever had happened in her past, this forgiving chicken now trusted and demonstrated a will to live that amazed even the avian vet. Ethologist Marc Bekoff (2007) commented that when studying animals, humans can only describe and explain their behaviour in words that make sense from a human-centred point of view, making anthropomorphism a powerful linguistic tool for those living with and caring for animals. For Anissa, who shared Zola's journey to heal, it was clear that Zola awoke each day happily clucking: "Let's do this human! Every day is a day closer to healing!"

Two weeks into her healing, Zola developed neurological issues that affected her balance and her confidence. Undaunted, Anissa set up an obstacle course so Zola could practice stepping up, down and over things. When the head spinning flustered her, she soon learned to respond to the words "heads up" to focus and get control of her body again. Using one of her 24 vocalisations, she would let Anissa know she was back on track and ready to get on with the serious business of scratching at the ground and enjoying normal chicken activities in safety.

Zola's unique personality, preferences and emotional capacity gradually emerged, and in a reversal of the norm, humans ran to oblige this tough little chicken. Hand feeding was her preference, but if the snack was not to her liking, Zola would decline with a distinctive sound that she seemed to know would be understood. Chickens use referential communication, which evokes mental representations of the causal event in the minds of receivers (Marino, 2017), a technique Zola perfected with her human flock. Fear of the nighttime attack never fully left her memory, but a red night light removed this stressor.

For two years, Zola enjoyed the life of a free chicken, playing in mud puddles, protected by flock members and humans alike, until her reproductive disease caught up with her and, despite treatment, she passed away. Zola had been destined to die as someone's dinner, a common outcome for discarded backyard chickens. Ironically, the strange advertisement offering this enticement saved her life and allowed Zola to live and die as a beloved individual whose unique personality and attitude had defied the odds. As all who care for rescued animals know, grief is inevitable and at times overwhelming, but it is also a reminder that these lucky few became someone, not something, and they died free of exploitation.

Figure 7.1 Zola survived a predator attack, and once she felt safe, lived every day to the fullest. (Image supplied by Mother Clucker's Microsanctuary and reproduced with their permission.)

FEELINGS AND EMOTIONS

Chickens are often perceived as having a low level of intelligence and lacking any of the psychological characteristics commonly associated with mammalian intelligence (Marino, 2017). This is despite evidence of their complex brain structure capable of processing and integrating information from the environment during decision-making. Within the skull bone are four ventricles, two of which are lateral and located within the telencephalon, covered by a thin cerebral cortex layer (Khamas et al., 2024). The telencephalon is also referred to as the cerebrum, especially in mammals where it is covered by a thicker cerebral cortex layer. The sparsity of this layer in birds—and chickens—prompted the erroneous assumption that birds had lesser brain power and were in fact "bird brains". Nothing could be further from the truth. In mammals, the brain's cortical surface includes raised folds or bumps (gyri) and indentations or grooves (sulci), which together increase the brain's surface and cognitive ability. Greater folding into grooves and ridges is associated with higher cognitive functions and problem-solving. As with many measures of perceived intelligence, humans were elevated to top of the list based on this measure. A bird's forebrain may lack the folds and grooves of mammals, but it is derived from the same neuro-anatomical substrate as the mammalian forebrain, indicating that chicken brains are certainly not lacking in functionality (Marino, 2017).

The third ventricle of the bird brain is located around the thalamus, which operates as an information relay point for sensory information to the telencephalon (Khamas et al., 2024). All four ventricles communicate in different ways, but there is no structure similar to the corpus callosum that connects the two hemispheres of a mammal brain. This makes birds natural "split brains" and raises the possibility that they may comprise a pair of conscious subjects, perhaps with even greater abilities than mammalian brains (Birch et al., 2020).

The chicken's smaller olfactory bulb compared to mammals is counter-balanced by larger optic nerves, suggesting that sight plays an important role in their lives. They have four colour photoreceptors in the eyes compared to humans' three, allowing them to see and be impacted by colours—such as ultraviolet light—that remain invisible to humans. Some research suggests that exposure to ultraviolent light may even be instrumental in improved feather condition and mediating the fear response, thus offering physiological and psychological welfare benefits (Wilcox et al., 2024).

Despite evidence strongly indicative of chickens being smarter than their stereotyped image, their emotional capabilities, both positive and negative, and their capacity to feel pain have also been overlooked at times. The lack of a mammal-like folded cerebral cortex, which is often held up as the gold standard of cognitive and emotional intelligence, may have skewed perceptions around chicken feelings. Starting with the conscious experience of pain, there is evidence that the chicken's dorsal pallium performs a similar function to the mammalian ganglia-cortico-thalamic loop (DeGrazia, 2020), facilitating the integration of motor, emotional, autonomic and cognitive responses to pain. This means birds have a suitable nervous system and brain for the conscious experience of pain, reinforced by other criteria for animal pain: nociceptive and endogenous opioid systems, protective behaviours and learned avoidance. It is impossible to ignore the fact that the chicken whose bones are breaking under their unnatural weight, or bare chested after plucking out their own feathers in frustrated stereotypies, or the chick who has been debeaked to mitigate severe feather pecking within a flock, are experiencing the same pain that any mammal—including humans—would feel.

Pain is often associated with fear, something chickens demonstrate in a range of different ways and circumstances. Capture and restraint can manifest as tonic immobility, while physiological responses such as increased heart rate and bodily fever in anticipation of a fearful stressor have been reported (Marino, 2017). For some chickens, fear may turn into traumatic memories and emotional scars, which can take time and patience to alleviate, even within the safety of a sanctuary as Cyana's story illustrates.

Cyana's story—Mother Clucker's Microsanctuary, USA

When the small black chicken arrived at Mother Clucker's Microsanctuary in 2019, she appeared sad, lost, grieving and terrified. These strong negative emotions escalated and coalesced into symptoms that in a human would be labelled as post-traumatic stress disorder (PTSD).

Humans are diagnosed with PTSD when repeated and unwanted recollections of a traumatic event make them relive the trauma over and over again. The fear and anxiety will not go away, leaving the person to live with a heightened sense of danger even when not at risk. Animals have played a large part in the treatment of human PTSD, whether with animal research models to develop anxiolytics and antidepressants, or the inclusion of animal-assisted therapy in the treatment regime (Insel & Landis, 2013). For decades, animal models of psychiatric disorders provided a powerful research tool for the identification of neural circuits, hormonal processes and behavioural mechanisms through which these disorders manifest, inferring that animals also have the neural substrates to develop psychiatric disorders in response to stressors in their own lives. However, this is rarely acknowledged or treated in commodified farmed animals.

Cyana was found injured on the side of a country road. A broken wing bone and other wounds suggested attack by a predator, but how she ended up on that road remained a mystery. Initially sent to a local animal shelter, Cyana proved unadoptable because her egg laying ability was no longer viable and she was deemed "useless". Relocation to Mother Clucker's transformed her into a somebody, but it soon became evident that Cyana's emotional healing would be long and painful.

Cyana was alone when rescued, a stressful situation for any animal who thrives on social connection. The return of her physical strength was accompanied by behavioural indicators of a grief response. Affective neuroscience positions grief as a primary emotion necessary to maintain social bonds (Panksepp, 2004). There is an urge to be reunited with companions after separation, and prevention of this can lead to separation distress and suffering. Cyana was slowly integrated into the Mother Clucker's flock, which seemed to allay some feelings of grief, but there were still occasions when her physical and emotional withdrawal threatened her slow recovery. This was most evident each evening before dusk when Cyana appeared to suffer a panic attack that left her isolated and screeching in extreme fear.

With symptoms reminiscent of PTSD, Cyana seemed unable to relinquish unwanted and intrusive memories of the predator attack, which most likely occurred in the fading light of dusk. The fear and anxiety threatened her ability to maintain connections with her new flock while still grieving the loss of her old group and so Anissa, from Mother Clucker's, devoted months to help this terrified chicken through her panic and grief.

Within days, Anissa recognised the need to carry Cyana to her sleeping quarters in the house well before dusk, and to use human vocalisations and tactile comforts to calm the stressed chicken. Stroking Cyana's neck and singing softly allowed Anissa's calming emotions to influence Cyana through inter-species emotional contagion. Daytime episodes of grief and panic were addressed in the same way as human and nonhuman animal bonded over past losses in a safe environment.

It took many months, but slowly Cyana's night anxiety showed signs of easing. Well before dusk each night, she would fly to the "communication log", a tall log at the door to the chicken run, to signal she was ready to be carried to bed. Once again demonstrating referential communication, all the chickens at Mother Clucker's learned that using specific vocalisations while standing on this log would generate the required human attention.

One year on, Cyana's PTSD and grief symptoms had faded, and her dormant personality emerged. Now the boss of the flock, her new-found confidence, curiosity and energy are testament to a resilient chicken who was allowed to heal on her own time and in her own way from a trauma that could just as easily have destroyed her. The bedtime ritual remains unchanged with Cyana's personal song, neck stroking, a kiss on the beak—but once done, Cyana's independence returns as she clucks loudly for peace and quiet among her newfound family. Chickens are often misperceived as lacking in psychological characteristics and cognitive ability, but Cyana did not subscribe to this stereotype. Identifying that human voices sometimes came out of a baby monitor set up to keep watch on the integration of a new flock member, Cyana would stare into the camera and voice her displeasure loudly. This display of object permanence (the ability to recognise that even out of sight, an object can still exist) is not evident in human infants for at least one year.

Human recovery from PTSD is a long and arduous journey that, for some, is never fully complete. Social support is crucial to the healing process, a need shared by humans and nonhuman animals alike. Both require patience, compassion, understanding and guidance to emotionally heal from past traumas and grief and regain feelings of safety. Unfortunately, the commodification of farmed animals means many will never have the opportunity to recover from day-to-day abuses and long-term traumas before being slaughtered to meet human needs. Cyana was one of the lucky few who received all the time she needed to recover within the safety of a farmed animal sanctuary.

Animal personality is no longer disputed among companion animals, but increasing evidence confirms that domestic farmed animals, including chickens, demonstrate unique characteristics related to exploration, activity, aggression, neophobia and fearfulness (Garnham & Løvlie, 2018). These traits show some degree of heritability, which has been capitalised on during selection for traits

Figure 7.2 Once Cyana regained her confidence, she became boss of the flock. (Image supplied by Mother Clucker's Microsanctuary and reproduced with their permission.)

suitable to mass confinement. Where a chicken's personality is reactive, they may find noxious stimuli more stressful than those with a proactive personality, while other personality types may be more likely to show unwanted behaviours such as feather pecking.

Chickens are smart and they love hanging out with their friends, passing the time scratching round and foraging. But they also have a huge emotional capacity if given the chance. Chickens show affective states and an awareness of the affective states of their companions. Sharing emotions through emotional contagion allows chickens to take social cues from each other and respond based on this rather than first-hand experience (Marino, 2017). Emotional contagion, a rudimentary form of empathy, is evident when chickens show negative responses to their companions being roughly handled. This advanced form of emotional capacity shows that they can become stressed without experiencing something directly, giving rise to potential welfare implications when thousands of chickens are confined in adverse conditions.

Basic empathy is also evident in the mother/infant bond, as demonstrated by Edgar and colleague's (2011) study. While a mother hen would show no emotional response to a puff of air in their cage, observing their chicks receiving this aversive puff of air was a different matter. Mother hens recorded physiological and behavioural changes, including increased heart rate, lower eye and comb temperatures, standing alert and maternal clucking. The researchers described this bodily reaction as a cognitively mediated response, having experienced the puff which did not stress them, but then translating this as being aversive to their chicks. The strength of the reaction depended on the mother's personality with more relaxed hens not showing a strong reaction. Taking the perspective of their chicks also shows perspective-taking, which has been associated with theory of mind in children (Garnham & Løvlie, 2018). Theory of mind is necessary when interpreting the thoughts, feelings and emotions of others, something the mother hens did when remembering the puff of air on their bodies and interpreting the impact this would have on the emotions of their chicks. Mother hens use a range of vocalisations with their chicks, especially cluck calls, to gather them around and set off to a new location. Broody hens will also talk to their unborn chicks in the last days before hatching with gentle clucks, and once hatched, interpret and respond to their chicks' two primary vocalisations: distress calls and pleasure calls (Gall et al., 2024).

Early studies on chicken welfare focused on their physical wellbeing—making a horrible situation a little less horrible—with less focus on the important things that make a chicken feel safe and happy. When chickens meet, they set up a dominance hierarchy, or pecking order, that involves pecking, jumping and clawing until dominance is asserted (Marino, 2017). Communication through vocalisations and different visual displays reinforces evidence of referential communication, whereby the chicken can evoke mental representations of the reason for their vocalisation in the mind of the receiver who responds appropriately. Males will tell females when they have found a yummy titbit, making movements of picking up and dropping the food as part of the complex social relationships among a group. Some cheeky males pretend there is food to attract the ladies, but the canny ladies soon work out the tricksters and stop responding to those males who call too often in the absence of food—using deception and counter-strategies similar to the most complex behaviours among mammals. Chickens can also discriminate between anticipation of a fear-worthy negative event and a beneficial positive event, with anxious pacing preceding an aversive event and preening and relaxation heralding a predicted enjoyable event.

Positive emotions can rarely be experienced by chickens in commercial farm settings during their short lives. However, once free of that system personalities, preferences, friendships and emotions can surface, as Lucky's story illustrates.

Lucky's new life—Mother Clucker's Microsanctuary, USA

Falling off the back of a truck does not sound lucky, but when the destination was most likely the slaughterhouse, that fall saved a small gold orange chicken's life. When Lucky and two of her sisters tumbled on to the road, only Lucky survived. Rescued by a lady following behind and delivered to Mother Clucker's Microsanctuary, Lucky's life trajectory was about to change dramatically.

Lucky's pelvis was broken, preventing her from standing or walking. Unfortunately, she also had severe osteoporosis (thinning bones) from excessive, unnatural egg laying. This confirmed Lucky was a spent hen, worn out from laying up to 300 eggs per year, then discarded and sent to slaughter when no longer productive.

Lucky entered the sanctuary a terrified and crippled little bird, dosed up on pain medications and supplements and put on crate rest to heal physically. Her reproductive system had shut down to facilitate healing as she progressed from bed rest, to a sling to facilitate standing, to being able to walk again within six weeks. Regular massages served a dual purpose: they kept her circulation moving while she remained immobile, and they encouraged her to feel safe and to trust humans, something that she had never experienced in her short life.

Chickens love to chatter with their 24 different vocalisations, and Lucky quickly learned to use her excited voice whenever Anissa from Mother Clucker's spent time with her. What she was saying did not matter—there was no longer fear or pain in the vocalisations that greeted Anissa.

A return to health brought with it a return to the unnatural egg laying regime selectively bred into laying chickens, but hormonal medication put a stop to that. Six months after commencing hormonal treatment, an X-ray showed that Lucky's bone density had improved without the constant drain on her body's calcium and minerals to create eggshells.

Lucky soon understood she had a voice that would be listened to, a privilege most caged or barn layer hens never experience. When she developed a strong attachment to Mike from Mother Clucker's, she learned his work schedule and would watch for him to pull into the driveway after work. Reserving a special chatter just for him as she quivered in anticipation, their mutual bond was unbreakable. A walk together, a quick snack, shoulder time and then a lift to bed left no doubt that Lucky's voice and preferences had been heard.

Lucky also discovered choice in her life—whether to take a dust bath or enjoy time loafing in the sun was up to her, although Lucky's gentle nature sometimes meant she had to wait for her pushier friends to vacate the dust bath. While Lucky would wait patiently, Cyana's more impatient personality (once she recovered from PTSD) saw her strutting around, clucking grumpily. Every chicken at Mother Clucker's has a unique personality and preferences that are respected, something that would have been impossible to predict for the young chicken who fell off a truck.

Figure 7.3 Lucky enjoys the sunshine and her new life. (Image supplied by Mother Clucker's Microsanctuary and reproduced with their permission.)

ETHICAL IMPLICATIONS

Often the study of emotions, especially in farmed animals, has focused on negative emotions and their avoidance as a welfare issue. However, as research broadened, it became evident that animals could experience positive feelings that were far more complex than simply the absence of negative emotions.

Many welfare assessment tools for chickens tend to focus on resource measures that address the quality of the environment, including access to feed, water, litter substrate, temperature and stocking density. These standard assessable measures however do not always offer insight into how the chicken is managing within that environment (Wilcox et al., 2024). Without factoring in a chicken's cognitive and emotional processes underpinning their abilities, interactions and perceptions, any welfare assessment is incomplete.

The story of Mrs Mangle and Tontine explains how even the most cruelly treated chickens can experience a positive effect in the right environment while continuing to suffer debilitating physical ailments. This story also shows the capacity of chickens to grieve and to be grieved.

Mrs Mangle and Tontine, a most unlikely friendship—Lefty's Place Farm Sanctuary, Australia

Who would have thought that an ex-sentinel chicken for the government, and a vitamin deficient ex-backyard chicken, could establish such an enduring bond in later life? This is the story of Mrs Mangle and Tontine, two disabled Isa Brown chickens whose legs would not work, but who loved each other dearly.

Sentinel chickens endure weekly blood tests to check for mosquito borne diseases and this was Mrs Mangle's life until being sent to live in a school. Tontine was dumped in a suburban front yard, unable to walk on legs rendered unusable from rickets (a skeletal disorder from vitamin deficiencies and lack of sunshine), only to suffer a fracture when thrown over the resident's side fence like garbage.

In 2023, Mrs Mangle arrived with a painful untreated leg injury at Lefty's Place Farm Sanctuary in the beautiful Macedon region of Victoria, Australia. Lefty's Place, run by photographer Tamara, has rescued, rehomed and given lifelong sanctuary to hundreds of animals, all of whom were recognised and respected as unique individuals, with life stories, personalities and future aspirations. The property is a paradise for chickens, with large enclosures and coops—and Tamara, who understands the unique needs of each resident on their journey to rescue, recovery and happy lives at Lefty's. Living with so many diverse chickens, Tamara can report with confidence that:

> People believe chickens don't feel the range of emotions we do, but I can tell you, they feel and express everything we do .... I hear all about their jealousies, their hates, their loves, their disagreements, their embarrassments and their whole range of emotions here every single day.
>
> *[Facebook, December 29, 2023. Reproduced with permission from Lefty's Place Farm Sanctuary.]*

When grumpy Mrs Mangle met sweet gentle Tontine, she spent three hours grumbling and yelling abuse at the interloper sleeping in her room. Undaunted, it was not long before Tontine was hobbling over from her side of the bedroom to crawl into Mrs Mangle's bed each night. And so began a most unlikely friendship between the cranky, rule-based Mrs Mangle and the sweet angel Tontine who earned the privilege of being the only chicken that Mrs Mangle ever groomed—or even liked. Renowned for regularly vocalising her complaints, chasing cockatoos off the property, demanding creamed corn and blueberries and generally being fat and spoilt, Mrs Mangle's façade crumbled in the face of the disabled newcomer.

Tontine was named after her love of the pillows that cushioned her broken pain-wracked body. Such was her start to life, Tontine never stopped worrying and sought refuge with the humans she trusted and with Mrs Mangle. Each night, she waited for Mrs Mangle to eat dinner together, chatting all the way though the evening and into the night. Barely able to walk, she would occasionally gather strength to hop a few steps in a desperate bid to be a normal little chicken doing normal chicken things. Feeling safe, cuddles, tomatoes, watermelon, oats, bio-cheese snacks, dust baths, sunshine, sushi hand-rolled treats and Taylor Swift's music brought joy to this chicken whose broken body and heart murmur were inevitably going to cut her life short.

In December 2024, Tontine ate her last sushi roll with joy, her eyes bright and her chatter enthusiastic. Shortly afterwards, her heart gave out and she died. Although devastated, Tamara recognised that the grief was not hers alone, as she explained:

> It's not just me who is feeling the tidal wave of grief, it's her best friend Mrs Mangle who has been hit with loss in the saddest of ways. She didn't eat for days and barely left her bed. As time goes on, Mrs Mangle will recover, her bond with that sweet hen was so special.
>
> *[Facebook, November 30, 2024. Reproduced with permission from Lefty's Place Farm Sanctuary.]*

The refusal to recognise grief in any animal other than humans shows an anthropocentric bias that is increasingly being eroded thanks to work in such diverse areas as ethology, affective neuroscience, ethics and philosophy. The question no longer centres on "Do animals grieve" but rather "How do animals grieve?" (Pierce, 2018), a troubling dilemma for those who continue to treat animals as unfeeling commodities for human use and abuse. Both Mrs Mangle and Tontine had been damaged and discarded. Relocation to Lefty's bestowed personhood on these unlikely friends but also introduced grief into the lives of Mrs Mangle, Tamara and all those who had the privilege of knowing Tontine as a unique individual.

Burying Tontine started the healing process for humans and chicken, and slowly, Mrs Mangle began to eat again. However, she continued to lose weight, and the vet was unable to find anything else wrong with her apart from old age and her deep-seated grief. She slept in completely different spots to the ones she had shared with Tontine. She mingled with the other chickens, but it was clear that they were not the ones she wanted to be with. Mrs Mangle continued to miss her little friend who was always so positive and joyful, even to the grumpiest hen at Lefty's.

Figure 7.4 Gentle Tontine looking up at her best friend, Mrs Mangle. (Image supplied by Tamara Keneally Photography, Lefty's Place and reproduced with permission.)

REFERENCES

Baldrey, V., Ragoonanan, K., & Bacon, H. (2024). Survey of the husbandry and biosecurity practices of backyard chicken keepers in the UK. *Veterinary Record, 195*(6), e4531. https://doi.org/10.1002/vetr.4531

Bekoff, M. (2007). *The emotional lives of animals: A leading scientist explores animal joy, sorrow, and empathy--and why they matter*. New World Library.

Birch, J., Schnell, A.K., & Clayton, N.S. (2020). Dimensions of animal consciousness. *Trends in Cognitive Sciences, 24*(10), 789–801. https://doi.org/10.1016/j.tics.2020.07.007

Bonnefous, C., Collin, A., Guilloteau, L.A., Guesdon, V., Filliat, C., Réhault-Godbert, S., Rodenburg, T.B., Tuyttens, F.A.M., Warin, L., Steenfeldt, S., Baldinger, L., Re, M., Ponzio, R., Zuliani, A., Venezia, P., Väre, M., Parrott, P., Walley, K., Niemi, J.K., & Leterrier, C. (2022). Welfare issues and potential solutions for laying hens in free range and organic production systems: A review based on literature and interviews. *Frontiers in Veterinary Science, 9,* 952922. https://doi.org/10.3389/fvets.2022.952922

CDC (2024). *2020–2024 Highlights in the history of Avian Influenza (Bird Flu) timeline.* https://www.cdc.gov/bird-flu/avian-timeline/2020s.html?CDC_AAref_Val=https://www.cdc.gov/flu/avianflu/timeline/avian-timeline-2020s.htm

Cosgrove, N. (2025). How many chickens are there in Australia? 2025 statistics. *PangoVet.* https://pangovet.com/statistics/how-many-chickens-are-there-statistics-australia/

Coulthard, S. (2022). *Fowl Play. A history of the chicken from dinosaur to dinner plate.* Head of Zeus.

DeGrazia, D. (2020). Sentience and consciousness as bases for attributing interests and moral status: Considering the evidence and speculating slightly beyond. In L.S.M. Johnson, A. Fenton, & A. Shriver (Eds.), *Neuroethics and nonhuman animals* (pp. 119–133). Springer.

Demarsh, B. (2024). Top Villain #5: Margaret Hudson of Burnbrae Farms. *Animal Justice.* https://animaljustice.ca/blog/margaret-hudson-of-burnbrae-farms

Edgar, J.L., Lowe, J.C., Paul, E.S, & Nicol, C.J. (2011). Avian maternal response to chick distress. *Proceedings of the Royal Society B, 278,* 3129–3134. https://doi.org/10.1098/rspb.2010.2701

Edison, R. & Esvelt, K.M. (2020). On mitigating the cruelty of natural selection through human genome editing. In L.S.M. Johnson, A. Fenton, & A. Shriver (Eds.), *Neuroethics and nonhuman animals* (pp. 17–32). Springer.

Ericsson, M. (2022). *Stress in chickens: Effects of domestication and early experience on behaviour and welfare* (1st ed.). Linkopings Universitet.

Ferreira, V.H.B., Guesdon, V., & Calandreau, L. (2021) How can the research on chicken cognition improve chicken welfare: A perspective review. *World's Poultry Science Journal, 77*(3), 679–698. https://doi.org/10.1080/00439339.2021.1924920

Fraser, D. & Weary, D.M. (2022). Applied animal behavior and animal welfare. In J.J. Bolhuis, L- A. Giraldeau, & J.A. Hogan (Eds.), *The behavior of animals,* 2nd edition*: Mechanisms, function, and evolution* (pp. 251–280). John Wiley & Sons, Inc.

Gall, G.E.C., Letherbarrow, M., Strandburg-Peshkin, A., Radford, A.N., & Madden, J.R. (2024). Exposure to calls before hatching affects the post-hatching behaviour of domestic chickens. *Royal Society Open Science, 11,* 240114. https://doi.org/10.1098/rsos.240114

Garnham, L., & Løvlie, H. (2018). Sophisticated fowl: The complex behaviour and cognitive skills of chickens and Red Junglefowl. *Behavioral Sciences, 8*(1), 13. https://doi.org/10.3390/bs8010013

Greene, R. (2025). 7 backyard chicken statistics to help you understand this growing trend. *Price of Meat*. https://priceofmeat.com/9550/backyard-chicken-statistics/#:~:text=Almost%2011%20million%20people%20in%20America%20raise%20chickens,the%20joys%20and%20rewards%20of%20backyard%20chicken%20keeping.

Grossblat, B. (2011). *Suicide food*. https://suicidefood.blogspot.com/

Havenstein, G.B., Ferket, P.R. & Qureshi, M.A. (2003). Growth, liveability, and feed conversion of 1957 versus 2001 broilers when fed representative 1957 and 2001 broiler diets. *Poultry Science*, *82*(10), 1500–1508. https://doi.org/10.1093/ps/82.10.1500

Henriksen, R., Johnsson, M., Andersson, L., Jensen, P., & Wright, D. (2016). The domesticated brain: Genetics of brain mass and brain structure in an avian species. *Scientific Reports*, *6*, 34031. https://doi.org/10.1038/srep34031

Insel, T.R. & Landis, S.C. (2013). Twenty-five years of progress: The view from NIMH and NINDS. *Neuron*, *80*, 561–567. https://doi.org/10.1016/j.neuron.2013.09.041

Jacobs, J. (1892). *English fairy tales*. G.P. Putnam's Sons.

Jones, M. (2014). Captivity in the context of a sanctuary for formerly farmed animals. In L. Gruen (Ed.). *The ethics of captivity* (pp. 90–101). Oxford University Press.

Khamas, W., Rutllant, J., & Smodlaka, H. (2024). Nervous system. In W. Khamas & J. Rutllant (Eds.), *Anatomy and histology of the domestic chicken* (pp. 171–182). https://doi.org/10.1002/9781119841739.ch12

Lefty's Place Farm Sanctuary (2023, December 29). *Facebook*. https://www.facebook.com/profile/100064382996343/search/?q=People%20believe%20chickens%20don%27t%20feel%20the%20range%20of%20emotions%20we%20do

Lefty's Place Farm Sanctuary (2024, November 30). Tontine. *Facebook*. https://www.facebook.com/profile/100064382996343/search/?q=Mrs%20Mangle%20who%20has%20been%20hit%20with%20loss%20in%20the%20saddest%20of%20ways

Marino, L. (2017). Thinking chickens: A review of cognition, emotion, and behavior in the domestic chicken. *Animal Cognition*, *20*, 127–147. https://doi.org/10.1007/s10071-016-1064-4

Mother Clucker's Microsanctuary (2025). Intro. *Facebook*. https://www.facebook.com/mothercluckersmicrosanctuary/

Ng, E. (2024 August 23). King adopts British Hen Welfare Trust's one millionth hen. *The Standard*. https://www.standard.co.uk/news/uk/highgrove-gardens-british-henrietta-king-queen-b1178090.html

Panksepp, J. (2004). *Affective neuroscience: The foundations of human and animal emotion*. Oxford University Press.

Pierce, J. (2018, August 24). Do animals experience grief? *Smithsonian Magazine*. https://www.smithsonianmag.com/science-nature/do-animals-experience-grief-180970124/#:~:text=A%20growing%20body%20of%20scientific,for%20or%20ritualize%20their%20dead.

RSPCA (2025). *How should I care for my backyard chickens?* https://kb.rspca.org.au/knowledge-base/how-should-i-care-for-my-backyard-chickens/

Siddiqui, S.A., Gerini, F., Ikram, A., Saeed, F., Feng, X., & Chen, Y. (2023). Rabbit meat—Production, consumption and consumers' attitudes and behavior. *Sustainability, 15*(3). https://doi.org/10.3390/su15032008

Singleton, D.A., Ball, C., Rennie, C., Coxon, C., Ganapathy, K., Jones, P.H., Welchman, D., & Tulloch, J.S. (2021). Backyard poultry cases in UK small animal practices: Demographics, health conditions and pharmaceutical prescriptions. *Veterinary Record, 188*, e71 https://doi.org/10.1002/vetr.71

Souvestre, M., Delpont, M., Guinat, C., Dumat, C., Guichard, L., Manis, L., Duret, H., Guérin, J-L., & Le Loc'h, G. (2021). Backyard poultry flocks in France: A diversity of owners and biosecurity practices. *Preventive Veterinary Medicine, 197*, 105511. ISSN 0167-5877. https://doi.org/10.1016/j.prevetmed.2021.105511

Wilcox, C.H., Sandilands, V., Mayasari, N., Yudha, I., & Anang A., & A. (2024) A literature review of broiler chicken welfare, husbandry, and assessment. *World's Poultry Science Journal, 80*(1), 3–32. https://doi.org/10.1080/00439339.2023.2264824

8

Roosters: Friends and protectors

INTRODUCTION

Growing up, I remember the daily joyful crowing of the rooster who lived a few houses down the street. Despite living in an urban area, Sir Rooster (as we called him) was an accepted and beloved member of our local neighbourhood. But times change, and today, Sir Rooster would be prohibited from this urban backyard—not for welfare reasons, but as perceived noise pollution in a suburban setting that can be overwhelming at times with human-made permissible noises. The exclusion of this reminder of the natural rhythm of day and night has had a far-reaching negative impact on those roosters who continue to be born into urban or commercial settings.

Roosters can be chatty characters beyond the morning crowing, with a range of vocalisations designed to both protect their flock and attract the ladies. There are alarm calls that differ when warning of aerial and terrestrial predators and the titbit calls to let their flock know of a tasty treat (Marino, 2017). In group settings, roosters work hard to attract the ladies and repel other males. Females gravitate towards those males who can provide the most food through courtship feeding and food displays, and even "eavesdrop" on the roosters when making their decision (Garnham & Løvlie, 2018). Some roosters have been caught out tricking the ladies with false information just to get them closer, but once the wily females work out the deception, they tend to avoid those tricksters.

Roosters demonstrate social awareness when adjusting their communication based on their audience. They have a greater tendency to alarm call when with familiar conspecifics and send out more alarm calls if they are closer to cover that they can run to or in the presence of a female (Marino, 2017). They even give more enthusiastic calls for better quality food but subdue their vocal food displays if dominant males are in the vicinity. Using the same call consistently for the same subject to invoke the same response suggests the advanced cognitive skill of referential communication. Referential communication relies on the rooster's ability to adopt perspective-taking when assuming the capacity of the

DOI: 10.1201/9781003518853-9

receivers to form mental representations without directly experiencing the event and respond appropriately (Garnham & Løvlie, 2018).

While the rooster's domestication timeline followed a similar trajectory to hens, their gender saw them used and abused in different ways within intensive animal agriculture, as well as human entertainment. In fact, humans have made the rooster's life increasingly difficult since the domestication of their reported ancestor, the Red Junglefowl (*Gallus gallus*), eight to ten thousand years ago.

SOME BACKGROUND HISTORY

For a while, domestication seemed to go well for roosters across a range of cultures and time periods when their primary role was to fertilise eggs and protect their flocks. Their history gradually became more colourful and diverse, ranging from being synonymous with male potency in medieval England, to a portent of disciple Peter's disownment of Jesus in the Christian bible, to being the tenth of the twelve animals in the Chinese zodiac (Coulthard, 2022). But somewhere along the way, things started to go horribly wrong.

Like hens, there can be multiple names for roosters depending on their age and country of residence. The term cockerel is used for a male chicken under one year of age, while rooster is a fully grown male chicken over 12 months. In some countries, roosters are referred to as "cocks", while a castrated male chicken is called a "capon".

The term cock is synonymous with many of the older forms of abusive entertainment involving roosters, including a particularly gruesome activity practised in England during the Christian Shrovetide. Shrovetide, called Carnival in some countries, is the week preceding Lent and perceived as the last chance for merriment before preparing spiritually for Lent's period of moderation and self-discipline. During this time, pleasure was juxtaposed with the need for sin confession in preparation for Lent, an ironic observation considering the pain and suffering inflicted on roosters during their forced participation to the death in some gruesome Lenten activities. During the 1400s, one popular traditional sport—called cock-threshing, or throwing at cocks, or cock-running—involved tying a live rooster by one leg to a stake or immovable object (Simpson & Roud, 2003). Players took it in turns to throw heavy sticks at the bird, with the winner being the one who made the final fatal throw. An alternative saw roosters placed in specially made earthen vessels with head and tail protruding, then suspended over three and a half metres (12 feet) above the ground while blindfolded players attempted to kill the bird with sticks. Children were allowed to play this game at Shrovetide, but its increased association with children and lower classes saw it give way to cock-fighting. These Lenten activities were some of the first of the blood sports to come under public pressure for abolition, and by the end of the eighteenth century, were rarely observed.

Cock-fighting, which had started during the Roman era and spread as the empire expanded, appealed to a higher-class spectator, including royalty, and would prove less amenable to abolition in the United Kingdom and around the world. In 1709, a book entitled *The Royal Pastime of Cock-fighting* by Robert Howlett and published in London by D. Brown clearly positioned cock-fighting as a royal—and profitable—occupation for humans, but certainly not for the roosters:

> The royal pastime of cock-fighting: or, the art of breeding, feeding, fighting, and curing cocks of the game; Published purely for the good, and benefit of all such as take delight in that royal, and warlike sport; To which is prefixed, a short treatise, wherein cocking is proved not only ancient and honourable, but also useful, and profitable.
>
> *(1709) by Robert Howlett.*

The book included a poem at the end in which the author presumed the cock's enjoyment and pride in being allowed to participate in this honourable sport. Lines such as "When first he [rooster] spies the Bloody trampled pit, He claps his wings, and Crows for Joy to See't [see it]" and "And when set down, he proudly struts along, [...] So eagerly to meet his threatening Foe" clearly position the rooster as a willing participant rather than a pawn in a lucrative type of entertainment.

To achieve the level of aggression required for this so-called sport, fighting cocks underwent extensive artificial selection that required more than aggressive behaviour (Budi et al., 2024). Prized physical attributes deemed crucial to fighting ability included large body size and muscular legs, scaly shins, and comb type. Fighting cocks came to symbolise social status as specific breeds emerged in a range of different countries. Champions also provided valuable income from stud fees, thus perpetuating the artificial selection of characteristics embodied in breeds such as China's Henan, Luxi, Tulufan, Xishuangbanna and Zhangzhou chickens; Japan's Shamo chickens and India's Kaunayen chickens. In Thailand, where the cock-fighting tradition dated back over 800 years, indigenous chickens were predominantly used as fighting cocks, including the Pradu Hang Dam, Lueng Hang Khao, Lao Pa Koi and Trad. Those deemed unsuitable were usually eaten, thus eliminating unwanted traits and strengthening the fighting traits. However, this also led to a loss of genetic biodiversity through in-breeding, with some potentially deleterious effects from extreme breed standardisation.

In the USA, cock-fighting generated considerable economic activity and was not relinquished without a fight. Wagering, entry fees to the fights and sale of fighting cocks made this blood sport a profitable business at the expense of those exhausted, injured or dead birds who were locked in a pit and forced to fight with steel blades tied to their legs. By 2021, cock-fighting was a crime in every state of the USA, despite the last two hold-out states, Louisiana and New Mexico, invoking the often-cited defence of "cultural prerogative" (Pacelle, 2021). Drawing on culture as a rationale to continue abusing roosters became a flawed rationale worldwide. Guamanian cock-fighters described it as part of their Pacific Island heritage, Vietnamese cock-fighters claimed it was in their blood, while some

Latino cock-fighters continued to assert it was their rightful tradition despite legal sanctions.

The countries of Latin America saw cock-fighting transition from this "most ancient animal sport" to prohibition and penalisation, with clandestine matches persisting in rural and remote areas (Mota-Rojas et al., 2023). For some communities, cock-fighting remained a source of employment and income, creating a status hierarchy among rooster owners and an attachment between rooster and human. In the short term, successful roosters received care and attention to maintain their extrinsic value as a source of revenue, but in the long term, this treatment would inevitably end when their winning streak ceased.

ROOSTERS IN CONTEMPORARY SOCIETY

In industrial farm settings, the life of a male chicken is often terminated within hours of hatching, raising the philosophical question of which environment provided greater meaning for the male of the species: the fighting cock obsessively cared for while winning but discarded when eventually losing, or the hatchling culled at birth? There is no right answer to this, especially given the way that some culling occurs.

Globally, tens of millions of male chicks in the egg industry are deemed useless and killed when newly hatched. Annual estimates from the United Kingdom put this figure at around 40–45 million male chicks that are incubated to hatching and then killed once their sex becomes evident (Gray, 2024). In the European Union, this figure escalates to around 330 million day-old male chicks in hatcheries being sorted and culled (Vinci, 2022), with a similar figure of 300 million in the USA (Compassion in World Farming, 2025). Globally, around six billion male chicks are culled annually. The female chicks' value lies in their ability to provide eggs, but the males are an undesirable and uneconomic by-product for both the lucrative egg industry and the meat industry. The conflicting requirements between fast-growing fattening regimes or increased laying performance resulted in specialised chicken lines selectively bred for growth rate or egg production. Male chicks born in the egg industry presented an economic liability as they grow too slowly and have limited thigh and breast muscle compared to the selectively bred meat chickens. As graphic images of the mass destruction of newborn male chicks circulated on social media and other media outlets, consumer outrage and advocacy escalated. A number of countries, including Germany and France, started banning routine culling, but globally, the practice continues in some regions.

Where culling remains legal, the death of so many tiny chicks is a horrible sight. Options include argon gas in a depleted oxygen chamber or carbon dioxide. Carbon dioxide raises ethical concerns related to the chicks' abnormal breathing patterns interpreted as air hunger and the activation of mucosal nociceptors resulting in pain and distress before death (Steiner et al., 2019). Maceration, the process of shovelling hundreds of fragile chick bodies towards rotating blades has attracted considerable public outrage for its gruesome bloodiness (Loeb, 2024).

Fluffy chicks, like many other young animals and infants, trigger an emotional "cute response" in humans. This innate mechanism evolved to protect the young and vulnerable and ensure their survival. Helpless, cute day-old chicks have all the ingredients to set this response in motion, especially when images of young chicks heading for maceration make their way to media outlets. If the blades are sharp and the process carried out competently, this visually distressing method has been described as faster and more humane than the prolonged suffering and death by carbon dioxide, thus providing a flawed justification for a process that is innately wrong (RSPCA, 2025).

Sexing of newborn chicks can be a distressing process in itself, whether done by colour in those breeds genetically modified so that males are coloured differently from females, or by opening the cloacal (anal vent) to check for genital organs in other breeds. Once hatched, the newborn chicks pass down a production line to be sexed and sorted at a rate of up to 1,000 chicks per hour, after which all the males and the weak or sick females are destroyed (Animal Welfare Committee, 2023).

Globally, countries that have banned the culling of day-old male chicks are the exception, with counter-arguments citing prohibitive costs of using pre-birth sexing methods, and the importance of culled chicks as foodstuffs for raptors, zoo species and exotic pets. This economic fact was pointed out in the 2023 Animal Welfare Committee report to the British, Scottish and Welsh governments with the warning that imported chick meat could be culled with lower welfare standards, as would the breeding of rodents to fill this niche.

Distress calling in newly hatched chicks indicates that they can experience negative states such as fear. The intensity of these calls match established behavioural and physiological measures of stress, and it can be modulated by anxiolytic drugs, providing substantive evidence that everything from sexing to gassing or maceration impacts the newly hatched male chick and the sickly female chicks (Animal Welfare Committee, 2023). The development of the central nervous system in a chick embryo commences on day 2 of incubation, followed by neurological sensory mechanisms in stages: tactile (day 6), proprioceptive-vestibular (day 8–10), taste (day 12), auditory (day 12–14), visual (day 18) and olfactory (day 20) prior to hatching on around day 21. Once neurons begin to work from around day 7 and brain waves in the embryo, as measured by electroencephalograph waves, are initiated by day 13, it is possible for the chick embryo to feel pain from this point to hatching, especially if precipitated by an invasive event.

A range of technologies to determine the embryo's sex at different times during incubation and with varying degrees of accuracy are currently available (Gray, 2024). These include allantoic fluid analysis by a range of different means, usually conducted after days 9–10, mass and fluorescence spectroscopy from day 4 of incubation, hyperspectral imaging to detect feather colour in brown layers from day 13 and magnetic resonance imaging from days 12 to 13 to view the embryo anatomy (Krautwald-Junghanns et al., 2018). Other technologies have been developed to influence the sex of the chick, including a combination of different sound frequencies and amplitudes combined with temperature and chamber gas concentrations to stimulate the degeneration of one testicle into an ovary, thus becoming functional females. While still in the experimental stage, it has been pointed out that genetic modification methods for altering chick embryo sex open the door to unintended negative side effects on the welfare of current and future progeny that may be difficult to reverse. In addition, these ovo sexing methods may not be affordable to small-scale breeders, thus promoting even larger intensive farming systems with associated welfare problems.

Creating dual-purpose economically competitive chickens by merging chickens bred for meat production (broiler line) and chickens bred for egg production (layer line) raised both technical and ethical challenges (Chuang et al., 2024). Dual-purpose hens lay smaller eggs, and cockerels yield less breast meat, but specific gene editing to rectify these issues could position these chickens as a genetically modified organism in consumers' eyes, thus alienating a portion of the commercial market.

Hatchling programmes

Male chicks face similar discrimination when born into school, nursery or care home chick hatchling programmes. These popular education tools seek to show younger children first-hand the creation of new life or allow older adults to reminisce about the chickens they may have once lived with. With an equal chance of the eggs being male, cockerels can end up with nowhere to go as they outgrow their enclosures and their function, and importantly, find their voice.

Schools and care homes may source their own fertilised eggs, or have them delivered by a hatchling programme supplier, along with an incubator, brooding box, food and bedding for the chicks. A few weeks after hatching, the supplier collects the hatchling kit along with any chicks that are no longer wanted or who have not been rehomed. During the weeks between hatching and return, chicks may become stressed if handled too much or too roughly, separated from other chicks or transported to and from children's homes. If the chicks are rehomed, that can be a problem for the 50 per cent who will inevitably turn out to be unproductive males. Unwelcome and illegal in urban areas, they can face the risk of euthanasia or abandonment (RSPCA, 2019). Hatchling programme suppliers promise to ethically rehome any returned chicks, but one supplier candidly acknowledged the potential difficulties with the males, commenting on their website that some of the boys may end up on the dinner table.

FEELINGS AND EMOTIONS

There seem to be few scenarios where male chickens can live out a long life in relative freedom—unless they meet up with a human who recognises them as unique individuals with feeling and emotions, and the right to life. For Mark, a dashing rooster living at Flow Farm and Sanctuary in China, this is exactly what happened.

Mark, a second chance—Flow Farm and Sanctuary, China

Mark was meant to be a female chicken and his life trajectory depended on that biological fact. When his male gender became evident, this created a significant dilemma for his owner who refused to let her beloved friend suffer just because he crowed and could not lay eggs. Now living harmoniously with his three lady chickens at Flow Farm and Sanctuary in China, Mark narrowly avoided the terminal fate inflicted on many male chickens around the world.

Flow Farm and Sanctuary is located in the heart of the Wenbishan Windmill Eco Park, a 270-acre nature reserve in Zhongshan, Guangdong Province, China (Flow Farm and Sanctuary, n.d.). The sanctuary evolved from a dream to create an experiential farm and animal sanctuary that provided both a home to rescued farm animals and a regenerative space for nonhuman animals, humans and nature to thrive in respect and harmony. Established in 2018 as an independent entity within the Windmill Eco Park, Flow Farm is now forever home to around 60 rescued farm animals including pigs, rabbits, chickens, ducks, ponies, mules, dogs and cats. Founder Justine Kwok and her team provide an environment where nonhuman animals are free to express their unique personalities while simultaneously advancing animal and environmental protection through experiential programmes. There is opportunity to encourage behaviour change in a new generation when young visitors interact with residents and learn respect, empathy and compassion for animals as individual sentient beings rather than commodities.

When an excited ten-year-old girl from Zhongshan, Guangdong Province, presented a tiny chick to her father, he was initially reluctant to allow the bird to remain in their apartment. However, the young girl's obvious attachment to the helpless little bundle of fluff swayed him and he finally agreed. The thought of fresh eggs was an added bonus to his daughter's happiness. The young girl raised the chick like any other companion animal, cradling the fluffy infant in her arms, taking her for walks on a leash and nurturing the mutual bond that grew stronger day by day. Everything changed when the chick uttered an unmistakable crow, and all thoughts of eggs and the neighbours' acceptance evaporated. The young girl's love for the growing chick, now renamed Mark, remained the same, but an apartment in an urban setting could no longer be home once his gender was confirmed.

The young girl's father scoured Zhongshan for someone to take in Mark, but things were looking grim as the shocking truth about roosters was reinforced with every knock-back. Mark, her beloved friend and companion, was worthless and a liability in society's eyes. Roosters do not lay eggs, and their meat is considered too "heated" for traditional Chinese medicine, which uses the warming and cooling nature of foods to balance the body's yin and yang and prevent and treat disease.

Finally, the young girl's father learned of the safety offered within Flow Farm and Sanctuary, but there was one more hurdle before Mark could join the other farmed animals. He would need a companion to maintain his wellbeing after separation from the young girl and his previous life. This did not daunt the young girl's father who rescued two female chickens destined for slaughter at the wet market and delivered Mark, Sophia and Lily to their new life at Flow Farm.

Sadly, Lily had long-term scissor beak, a congenital deformity in which the upper and bottom beaks grow in two directions and make it difficult for the chicken to pick up food. Lily died soon after arriving at Flow Farm, leaving Mark and Sophia to enjoy their large enclosure and the opportunity to finally express their natural, species-specific behaviours. Daily dust baths to feel fresh and clean were a luxury they never grew tired of as their physical and emotional wellbeing peaked.

When puberty hit Mark, the dynamics changed as poor Sophia became increasingly traumatised by Mark's insatiable young man appetite. As a "no purchase" sanctuary, the decision was made to separate young Mark from the object of his affections and wait it out. Sure enough, it did not take long before two British volunteers rescued four female chickens crammed into a rusty cage and just minutes away from slaughter. After a month's quarantine and mounting excitement, Mathilda, Dorothy, Rose and Gigi met Mark who welcomed them with open wings. Sophia and Mathilda enjoyed another year of freedom before passing away, but Mark, Rose, Gigi and Dorothy continued to happily share the largest territory ratio-wise in the sanctuary.

Roosters and chickens have excellent long-term memories, and Mark was no exception. With maturity and a flock to protect, Mark became dominant and territorial of his kingdom, often refusing entry to strangers or infrequent visitors. But there is one exception. Whenever the young girl visits, you will see Mark race towards her and fling himself onto his favourite rock to greet her. As the young girl softly speaks to him, it is clear that this dominant protector of his ladies has never forgotten the love and cuddles he once shared with the young girl. Their loving bond remains intact providing heart-warming evidence that roosters are just as sentient and sentimental as humans—if given the opportunity to live.

All rooster flocks

If still living with their mothers, chicks naturally start to become independent from their mothers from around five to six weeks. They also start to form social hierarchies that remain relatively stable. In a mixed setting, sexually mature males generally rank higher than females (Garnham & Løvlie, 2018). A prevailing rooster myth is that they are naturally aggressive towards each other, a stereotype reinforced by images of cock-fighting. In reality, each rooster has their own traits and history that will shape their personality and the way they interact. Some rescued cock-fighters may have difficulty integrating due to the way humans have bred and trained them, while others have a history rooted in severe trauma and may appear confrontational as a defence mechanism.

Figure 8.1 Mark stands on a rock with his ladies around him. (Image supplied by Flow Farm and Sanctuary and reproduced with their permission.)

There is another long-held assumption that roosters cannot co-exist with each other and the most natural social arrangement sees one rooster being head of a flock of hens. This has been challenged at a number of farmed animal sanctuaries, as well as the same-sex groupings, especially of juvenile roosters, found in natural settings among the chickens' early relative, the Red Junglefowl. This disruption of gender stereotypes suggests it may be the humans who struggle with the concept of non-normative social groupings rather than the roosters themselves (Rosenfeld, 2022).

Previously named "bachelor rooster flocks", the inaccurate connotations within this label have been increasingly critiqued. For example, Triangle Chicken Advocates (2019) comment that members of their all-rooster flocks fail to meet the traditional criteria for bachelors. They are not lacking deep and meaningful relationships with their fellow roosters in the pack, making their connections just as important and valid as a rooster and a hen pack. With many roosters culled or abandoned because of false stereotypes, all-rooster flocks provide an option for rescue from slaughter or forced illegal cock-fighting.

However, it is necessary to remove female chickens from sight to mitigate potential conflict from fighting over females as a perceived resource. Remove this resource from the equation, and males will develop their own hierarchy, following the same processes of establishing status and social structure as in a mixed-gender flock (Rosenfeld, 2022). A dominant male will become head of the flock, creating an environment where strong bonds can be formed between members, and he will work to order the others and protect them from danger. Flock members form friendships, watch out for each other, alert each other to potential dangers, dust bath together and snuggle at night, no different to a mixed-gender flock (Triangle Chicken Advocates, 2019). At Black Sheep Animal Sanctuary Ōtaki in New Zealand, their all-rooster pack shows just how well this natural system can work.

Friendship—the all-rooster pack and Black Sheep Animal Sanctuary Ōtaki, New Zealand

An hour's drive north of Wellington, Aotearoa New Zealand, lies a 64-hectare (26 acre) haven for rescued farm animals, including ex-dairy cows, ex-factory farm pigs, ex-battery chickens, goats, ducks, roosters and more. Black Sheep Animal Sanctuary Ōtaki provides a refuge for abused, neglected and mistreated animals where they can live out their lives with dignity, care and respect in the company of others of their kind. Since 2009, the Sanctuary, owned by the New Zealand Animal Protection Society, has promoted animal rights and animal liberation, opposing all forms of oppression including sexism, inequality and environmental destruction (Black Sheep Animal Sanctuary Ōtaki, n.d.).

At Black Sheep, every animal is an individual in life and death, something they could never have achieved when anonymous commodities for human consumption within the industrial farm system. In life, they experience the basic right to express natural, species-specific behaviours, so essential for their emotional wellbeing. In death, they are buried at the far edge of the goat paddock with a tree planted above them as a memorial. They are missed and remembered, as the loving, unique, sometimes grumpy, sentient animals who had the opportunity to live and die as loved individuals.

Walk down the hill at Black Sheep Sanctuary, past the rescued resident cows, and you will come to rooster orchard where a special group of all-roosters live their best lives. There are no female chickens at Black Sheep, a decision made based on the high number of calls from people seeking to rehome the roosters they could no longer accommodate at their current settings. Roosters are notoriously hard to rehome for a number of reasons: false stereotypes about their aggression, concerns about their noise levels, legal constraints or simply because they are perceived as non-productive for human needs. Nothing could be further from the truth. Dating way back to their ancestor, the Red Jungle Fowl, roosters have evolved to be protective, caring, inquisitive and cheeky, with a flow-on effect of heightened awareness of dangers and predators.

However, if housed with female chickens, the rooster's protectiveness of his ladies can extend to maintaining the established pecking order within the flock, thus making it difficult to introduce mature roosters into the sanctuary. In order to help as many roosters needing a forever home as possible, the decision was made to focus on all-rooster flocks—with the added bonus of being privy to a completely different view of how roosters behave when not competing over the ladies.

Danny from Black Sheep provided some wonderful anecdotes arising from their decision to provide a permanent home to these unwanted roosters. Roosters crow in the morning, throughout the day, and at night for many reasons, including some reasons known only to themselves. They might take it in turns to crow in their established pecking order or social rank or, as Danny describes, they are simply telling everyone "Hey everything is okay over here" and then another will shout back "Yep everything over here okay too".

Within the rooster flocks, the group dynamics can change over time as friendships emerge among pairs and trios of more closely bonded individuals. At Black Sheep, these friendships, instigated by the roosters themselves, include Mr Floofy Pants and Chips, Goldie and Glowie, and Cruella and Einstein, all of whom use around 24 different vocalisations to communicate with each other within the larger flock. Finding a treat is especially exciting as one will sing a lovely song to let their friend know they have found something great and they are keen to share it. Sadly, when a best friend dies, the other is left to grieve as happened to Matata when his best friend Spud died in 2024. Spud now appears on Black Sheep's Rest in Peace website page, a testament to how every life leaves a lasting imprint on animal and human hearts alike at the sanctuary.

Roosters also learn from observing and mimicking each other, which in human psychology is called social learning. Psychologist Albert Bandura's (1977) social learning theory relies on a mix of observing and imitating the behaviours, attitudes, and emotional reactions of others. Danny described how the roosters watch and learn whenever a new enrichment toy is introduced or when they first used the auto feeder. Once one rooster worked out what to do in the novel situation, their friend soon acquired the new response simply from watching. The complexity of forming a mental representation of the behaviour and symbolically storing the behaviour in their mind until they had the opportunity to successfully imitate it are quite sophisticated actions for an animal who is often discarded as being useless or noisy.

ETHICAL IMPLICATIONS

Stereotypes, misinformation and economic realities in intensive animal agriculture can make life difficult—and short—for chicks who happen to be born male. Descriptors, such as noisy, non-productive, aggressive or illegal, and ignorance as to what to do with them, have created an urgent welfare issue for unwanted chicks and roosters. There is a need for targeted education campaigns to counter their negative reputation and provide information on care for roosters compared to hens. While unable to offer sanctuary to roosters, the British Hen Welfare Trust, a charity that saves and rehomes commercial laying hens throughout the United Kingdom, set up an online "Lonely Hearts Club for Cockerels" to help roosters find a new home in safety (British Hen Welfare Trust, 2025).

Figure 8.2 Mr Floofypants and Chips have chosen to be friends within the all-rooster pack. (Image supplied by Black Sheep Animal Sanctuary Ōtaki and reproduced with their permission.)

However, globally many newly hatched male chicks continue to be killed or abandoned. All-rooster flocks offer an alternative for the lucky few, but the wider problem remains. The economic pressures driving the culling of day-old male chicks have seen most welfare debates focus on the most humane way to kill them rather than the larger ethical issue of mass killing of sentient beings who do not meet human requirements.

Figure 8.3 Cruella and Einstein, another pair who formed a friendship within the all-rooster pack. (Image supplied by Black Sheep Animal Sanctuary Ōtaki and reproduced with their permission.)

REFERENCES

Animal Welfare Committee (2023). *Opinion on alternatives to culling newly hatched chicks in the egg and poultry industries*. https://www.gov.wales/sites/default/files/publications/2023-11/awc-opinion-alternatives-culling-newly-hatched-chicks-poultry-industry.pdf

Bandura, A. (1977). *Social Learning Theory*. Prentice Hall.
Black Sheep Animal Sanctuary Ōtaki (n.d.). About us. The Black Sheep Animal Sanctuary Ōtaki. https://www.theblacksheep.org.nz/
British Hen Welfare Trust (2025). *Cockerel adoption*. https://www.bhwt.org.uk/cockerel-adoption/
Budi, T., Luu, A.H., Singchat, W., Wongloet, W., Rey, J., Kumnan, N., Chalermwong, P., Nguyen, C.P.T., Panthum, T., Tanglertpaibul, N., Thong, T., Ali, H., Vangnai, K., Chaiyes, A., Yokthongwattana, C., Sinthuvanich, C., Han, K., Antunes, A., Muangmai, N., Duengkae, P., & Srikulnath, K. (2024). Purposive breeding strategies drive genetic differentiation in Thai fighting cock breeds. *Genes Genomics*, *46*(10), 1225–1237. https://doi.org/10.1007/s13258-024-01561-3
Chuang, E.Y.S., Wellmann, R., Meijboom, F.L.B., Tetens, J., & Bennewitz, J. (2024). Simulation of dual-purpose chicken breeding programs implementing gene editing. *Genetics Selection Evolution*, *56*(1), 7. https://doi.org/10.1186/s12711-023-00874-3
Compassion in World Farming (2025). *EggTrack 2024 – U.S. & Canada: Pushing for progress in egg-laying hen welfare*. https://www.ciwf.com/food-business/eggtrack/eggtrack-usa-report/#References
Coulthard, S. (2022). *Fowl play. A history of the chicken from dinosaur to dinner plate*. Head of Zeus.
Flow Farm and Sanctuary (n.d.). *Welcome to Flow Farm and Sanctuary*. https://farmflow.org/
Garnham, L., & Løvlie, H. (2018). Sophisticated fowl: The complex behaviour and cognitive skills of chickens and Red Junglefowl. *Behavioral Sciences*, *8*(1), 13. https://doi.org/10.3390/bs8010013
Gray, A. (2024). Call for ban on mass culling of male chicks. *Veterinary Record*, *194*(2), 58. https://doi.org/10.1002/vetr.3863
Howlett, R. (1709). *The royal pastime of cock-fighting*. D. Brown. https://archive.org/details/the-royal-pastime-of-cock-fighting-1709
Krautwald-Junghanns, M.E., Cramer, K., Fischer, B., Förster, A., Galli, R., Kremer, F., Mapesa, E.U., Meissner, S., Preisinger, R., Preusse, G., Schnabel, C., Steiner, G., & Bartels, T. (2018). Current approaches to avoid the culling of day-old male chicks in the layer industry, with special reference to spectroscopic methods. *Poultry Science*, *97*(3), 749–757. https://doi.org/10.3382/ps/pex389
Loeb, J. (2024). Day-old chick slaughter 'not inhumane'. *Veterinary Record*, *194*(7), 243. https://doi.org/10.1002/vetr.4094
Marino, L. (2017). Thinking chickens: A review of cognition, emotion, and behavior in the domestic chicken. *Animal Cognition*, *20*, 127–147. https://doi.org/10.1007/s10071-016-1064-4
Mota-Rojas, D., Strappini, A., Whittaker, A.L., Ghezzi, M., Titto, C.G., Calderón-Maldonado, N., Mora-Medina, P., Domínguez-Oliva, A., Gómez-Prado, J., Hernández-Ávalos, I., & José-Pérez, N. (2023). Controversial topics in animal welfare in Latin America: A Focus on the

legislation surrounding the human-companion animal relationship and animals used for recreational practices. *Animals, 13*, 1463. https://doi.org/10.3390/ ani13091463

Pacelle, W. (2021). The sun finally sets on legal cockfighting in the United States. *Center for a Humane Economy*. https://centerforahumaneeconomy.org/2021/10/19/the-sun-finally-sets-on-legal-cockfighting-in-the-united-states

Rosenfeld, H. (2022) Sanctuaries as multispecies safe spaces. In E. Cudworth, R. E. McKie, & E. Turgoose (Eds.), *Feminist animal studies: Theories, practices, politics* (pp. 165–182). Routledge.

RSPCA (2019). What are the animal welfare issues with chick hatching in schools? *RSPCA Knowledgebase*. https://kb.rspca.org.au/knowledge-base/what-are-the-animal-welfare-issues-with-chick-hatching-in-schools/

RSPCA (2025). What happens with male chicks in the egg industry? *RSPCA Knowledgebase*. https://kb.rspca.org.au/knowledge-base/what-happens-with-male-chicks-in-the-egg-industry/

Simpson, J. & Roud, S. (2003). *A dictionary of English folklore*. Oxford University Press. Print ISBN-13:9780198607663. https://www.oxfordreference.com/display/10.1093/acref/9780198607663.001.0001/acref-9780198607663-e-212

Steiner, A. R., Flammer, S. A., Beausoleil, N. J., Berg, C., Bettschart-Wolfensberger, R., Pinillos, R. G., Golledge, H. D. W., Marahrens, M., Meyer, R., Schnitzer, T., Toscano, M. J., Turner, P. V., Weary, D. M., & Gent, T. C. (2019). Humanely ending the life of animals: Research priorities to identify alternatives to Carbon Dioxide. *Animals*, *9*(11), 911. https://doi.org/10.3390/ani9110911

Triangle Chicken Advocates. (2019) *A quick note on rooster flocks* [Facebook]. Available at: www.facebook.com/475783969194501/posts/1944260052346878

Vinci, C. (2022). Male chick culling. *European Parliamentary Research Service*. https://www.europarl.europa.eu/thinktank/en/document/EPRS_ATA(2022)739246

9

Turkeys: Unique individuals, not celebratory meals

INTRODUCTION

When checking the meaning of the word turkey, the Cambridge dictionary (2025) lists several definitions, including something that fails badly, a stupid or silly person, a large bird grown for its meat on farms and the flesh of this bird used as food. How sad that this majestic, loving and cheeky bird has been minimised to folly or food, when anyone who knows a turkey can attest to their unique personalities and loving nature. Preena, Penelope and Colin, Susan, Barbara and Mr Biggles, whose stories appear later in this chapter, all show that turkeys are so much more than meat or stupidity.

With the scientific name of *Meleagris gallopavo* bestowed by Swedish naturalist Carl Linnaeus in the 1758 *Systema Naturae* (10th edition), it remains unclear as to how these quirky birds obtained their common name of turkey (Brant, 1998). The first Europeans to sight them in America believed they were a type of Guinea fowl. As Turkey and Arabian territories provided a gateway for exotic imports to Europe in the Middle Ages, some believe they were first referred to as the bird from Turkey and later shortened to Turkey or Turkey fowl. Others comment that the French name for the turkey translates into English as "from India", while the Indigenous Americans in this wild bird's homeland had over 100 names for them. For some reason, the name turkey stuck, despite some more appropriate homeland names.

SOME BACKGROUND HISTORY

For millions of years, wild turkeys roamed North America. Bones found in the La Brea tar pits of Los Angeles in California date back to the Pleistocene era, over 10 million years ago (Brant, 1998). When European explorers arrived in North America and Mexico in the 1500s, they not only discovered a new continent, but also an abundance of wild turkeys. Christopher Columbus fed on turkey meat

DOI: 10.1201/9781003518853-10

when he landed on the coast of Honduras in 1502, while Cortez found large numbers of turkeys in Mexico in 1519.

Their journey to Spain may have been by Columbus or Cortez, or both, but they continued to be sent over and spread rapidly throughout Europe. There were turkeys in Spain and Germany by 1530 and in England by 1541, appearing in the famous diaries of Samuel Pepys with reports of large flocks being driven from surrounding counties to London markets (Brant, 1998). Relentless hunting and displacement made the wild turkey almost extinct in some areas of North America and Mexico, to be replaced by the domesticated turkeys that returned with the English colonists in the 1600s.

Meanwhile, back in North America, a difficult first winter in a strange land saw early settlers in Plymouth, Massachusetts set a precedent that would have fatal implications for turkeys right up to the present day. In 1623, after a healthy harvest, settlers celebrated with a three-day feast to give thanks—with the bountiful wild turkeys providing a centrepiece to these celebrations. And so began the turkey's key role in the thanksgiving holiday every November that would eventually culminate in around 46 million domesticated turkeys raised to be slaughtered each Thanksgiving in the USA (Farm Sanctuary, 2025). When combined with the traditional Christmas turkey, the total reaches a staggering 200 million turkeys in the USA each year, with over 90 per cent of them raised in industrial farm systems and kept in severely overcrowded conditions. Canadian statistics report 2.7 million whole turkeys purchased for Thanksgiving, and over six million for the whole year (Rink, 2023), while in the United Kingdom, an estimated ten million turkeys are killed at Christmas (Sparks, 2024), and around five million in Australia, also mostly at Christmas (Farm Transparency Project, 2020).

Thanksgiving turkeys are a legacy of the North American early settlers in the 1600s, but Christmas turkeys had a different origin. There was no religious or festive association between turkeys and Christmas; they simply became fashionable after King Henry VIII in England swapped his usual goose for a turkey in the mid-1500s (Lally, 2024). Turkeys became the must-have for Christmas festivities with Queen Victoria enjoying turkey at Christmas and Charles Dickens concluding the book *A Christmas Carol* with Ebenezer Scrooge giving the Cratchit family a turkey (Minney, 2022). It helped that the turkey could be rapidly fattened prior to Thanksgiving or Christmas, and once this tradition was established in a growing number of countries, the turkey's fate was sealed.

By 2023, over 250 million domesticated turkeys were being kept globally (Shahbandeh, 2025), with the majority confined in industrial farm systems (Ritchie, 2024). Compared to their free-flying wild relatives, these turkeys were selectively bred with bodies capable of providing considerably more meat for human consumption at the end of their shortened lives. They had also become the second most important source of poultry meat in the world (Walk et al., 2023).

TURKEYS IN CONTEMPORARY SOCIETY

Domesticated turkeys look quite different to their wild relatives. In the early twentieth century, an English financier, Lord Rothschild, had the idea to gift extra-large turkeys to his friends and colleagues at Christmas (Brant, 1998). Rothschild was prepared to pay a premium for these birds and tasked a turkey farmer to create and supply these larger than usual birds. The farmer collected the largest birds he could find from breeders all around the United Kingdom, thus creating his own strain he called the Sheffield Bronze which saw males weighing in at 18 kilograms (40 pounds) and females at 9 kilograms (20 pounds) at maturity. In contrast, male wild turkeys weighed around 5–11 kilograms (11–24 pounds) and the females 2–5 kilograms (5–12 pounds).

After decades of selective breeding, genetic alterations finally produced turkeys double the size of those in the 1960s, with resultant injuries and death from struggling to cope with their enormous body weight. Not only were they bigger than their wild relatives, domestic turkeys grew three times faster and had 2.8 times larger pectoralis major muscle mass compared to wild turkey strains (Walk et al., 2023). By the time they reached market weight for slaughter, many suffered from deformities and heart conditions, with some unable to reach their food and water. They could not fly and struggled to walk on legs that buckled and broke under their massive unnatural weight, a far cry from their wild relatives with running speeds of up to 40 kilometres per hour (25 miles per hour) and flying speeds up to 88 kilometres per hour (55 miles per hour). Even if they had retained a fraction of these abilities, the growing industrial farm systems denied them the opportunity to run or fly when housed with thousands of other turkeys awaiting their inevitable slaughter for human celebratory feasts.

Increased size was not enough though, and the next consumer-driven change related to feather colour and a push to convert many commercial turkeys to white feathers rather than the usual darker plumage (Brant, 1998). At some point, turkey producers became convinced that consumers did not like the dark spots of black pigment left in feather follicles from the turkeys' naturally dark plumage. The breast area in particular was felt to be unsightly and therefore impacting quality. White feathered turkeys left no ugly spots of black pigment, thus motivating many turkey breeders to transition their production from the broad breasted Sheffield Bronze to the Broad Breasted White. With white feathering a single recessive gene, once established the white trait breeds did not spontaneously produce any dark offspring.

As turkeys found themselves in larger and larger groups to meet human demand, their highly competitive social system and behaviours to establish and maintain social hierarchies were thrown into chaos (Marchewka et al., 2013). Among wild turkeys, group hierarchies are established through close kin relationships between relatives. External males faced aggressive fights and rejection from the closed units that tended to be created for life. As group size and space availability increasingly changed with domestication and industrialisation, there was an urgent need to learn more about turkey behaviours for both welfare and economic reasons.

Turkeys need fattening for longer than chickens, with hens typically reared for 12 weeks and males (toms) for up to 20 weeks (Marchewka et al., 2019). In their natural habitat, turkeys can live up to ten years, showing the devastating impact of human intervention. These varying lengths of the production cycle of turkeys meant males and females needed dividing, often by a partition, but still housed in the open space of the same building.

Not surprisingly, insufficient space, inadequate housing and poor management conditions led to greater risk of broken wings when birds hit the pen walls or other birds during aggressive encounters, or received injurious pecking when unfamiliar members were introduced to the group. Severe head and feather pecking could cause tissue damage and mortality, resulting in compromised welfare and, for producers, negative economic implications (Bartels et al., 2020). Enclosure size seemed to be the determining factor, especially as commercial flocks reached numbers in the thousands and prohibited any form of hierarchical system to maintain order. Wild turkeys living in natural settings occupy themselves with foraging and fighting for a reason, be it mates or territory. Fights among wild turkeys can be common, but not to the injurious and long-lasting level of confined domestic commercialised turkeys who cannot spatially avoid each other nor establish a hierarchy as a pre-requisite for cooperative living and courtship.

As stocking density created ever-increasing issues—including head and back wounds, vent wounds, leg and foot problems, breast blisters, air sac disease and feather plucking—beak trimming became a cruel, but legal option in many countries. Within days of hatching, the end of the tiny poult's sensitive, nerve-filled beak was burned or cut off with no form of pain relief. According to the Humane League of United Kingdom, these birds showed behavioural indicators of depressed mood and long-term chronic pain (Sparks, 2024), while a report from the European Food Safety Authority (EFSA, 2024) described potential difficulties in feeding, drinking or foraging immediately after beak trimming, and the risk of lifelong suffering through beak deformation and chronic pain. Of concern, studies among other poultry suggested that beak trimming could increase feather pecking when frustration at not being able to grasp feathers prompted greater motivation to continue feather pecking and pulling until the action was complete. Feather pecking could also spread around the flock through social learning when observing other turkeys performing this action. Given the near-impossibility of keeping non-beak-trimmed fattening turkeys under extremely high-density conditions without pecking, the logical response was to reduce densities—and preferably cease farming turkeys altogether—but this was largely ignored as an option.

The EFSA (2024) report also identified toe trimming as another negative procedure performed on young chicks. Defined as the burning of the germinal bed of the three forward facing digits on turkeys, this cruel procedure sought to prevent injuries and scratches among fattening turkeys but also protect handlers of turkey breeders when harvesting the semen for artificial insemination. Usually performed without pain relief, turkeys experienced acute then chronic pain, and subsequent difficulties walking, perching and foraging (Farm Sanctuary, 2025). Any turkey suffering from the aftermath of beak and toe trimming was likely destined for early slaughter, unless they had the good fortune to end up at a farmed animal sanctuary.

Preena, becoming "someone"—Edgar's Mission, Australia

When Preena and nine chirping poults arrived at Edgar's Mission in Victoria, Australia, their trimmed beaks and amputated toes painted a grim picture of where they had likely come from. One week short of Christmas 2017, the Edgar's Mission team was excited to learn of a "gift" awaiting them at their front gate. Imagine their surprise when ten curious little individuals poked their heads out of the box. In Australia, an estimated five million turkeys are slaughtered each Christmas to satisfy human celebratory needs. Thanks to a stranger's act of kindness, that year there were ten less Broad Breasted White turkeys served up on a plate. Instead, as the tiny bug-eyed infants were lifted gently from the box and settled into a bed of soft straw, they immediately became "someone" with a future at Edgar's Mission.

Preena and her friends were safe from slaughter, but they could not escape the genetic manipulations that had doubled the size of turkeys since the 1960s so people could enjoy more breast meat on their plate. Left with a legacy of legs that risked buckling—or breaking—under their unnaturally fast growth, most turkeys would never experience the simple pleasures of walking and foraging during their few short months of life before slaughter.

Almost eight years after that fateful Christmas, only Preena remained, her stubby toes and blunted beak a haunting reminder of the callousness with which they were removed in her previous life. Her nine turkey friends enjoyed love, kindness, freedom, fun and curiosity—topped off with healthy servings of watermelon—during their time at Edgar's Mission, and all lived longer than the industry-dictated three to four months of miserable life.

Preena loved people, striding over to friends and first-time acquaintances alike to nestle her body to the ground, tuck her head under and invite a gentle caress of the soft, caruncle skin on her head and neck. When famous ethologist and animal behaviourist Marc Bekoff (2024) commented that the very presence of animals, with their free-flowing emotions and empathy, is critical to human wellbeing, he could easily have been talking about Preena when she purred like a cat at this attention. This simple and natural act of happiness over something so simple from an animal still bearing the legacy of humanity's cruelty was humbling and contagious.

As sole survivor of the ten tiny chirping turkey poults, Preena commanded the team at Edgar's Mission with grace and grandeur. When summer heat peaked around Christmas time in Australia, Preena enjoyed a chauffeur-driven ride to the air-conditioned nursery on sweltering days. Burdened by an excessive and

Figure 9.1 Preena checks out what is happening at Edgar's Mission, her stubby toes a reminder of the life from which she escaped. (Image supplied by Edgar's Mission and reproduced with their permission.)

unnatural muscle mass, Preena struggled with heat stress, heart problems, mobility issues and musculoskeletal strain, while still maintaining her friendly, cheeky and happy demeanour that made working innovatively to accommodate Preena's needs a pleasure. Preena wriggled her way into the hearts of all who met her but remained a stark reminder that there are still millions of turkeys who will never have the opportunity to experience love, respect and freedom before they are served up on a plate.

In August 2025 Preena passed away, surrounded by soft words and the gentle hands of those who loved her.

Where a turkey lives can determine how much room they have to move as there are no consistent global stocking density standards. Even within one country, requirements can be open to some interpretation, making the turkey's life dependent on the producer's understanding and enactment.

Best practices for turkey management in Europe specify: "The ability of the birds to move and behave normally (pecking and wing flapping possible)" and state a maximum stocking density of a turkey house not exceeding 60 kilograms per metre square (132 pounds/10.8 square feet) (AVEC, 2024). In the USA, Standards of Conduct state that turkeys must be: "Free from cages with room to strut about within protective barns or with access to the outdoors" (National Turkey Federation, 2025), while Canada requires: "Birds must have enough space to move freely and be able to stand normally, turn around, and stretch their wings without difficulty. Space allowance must be sufficient to allow all birds to be able to sit at the same time" with a list of maximum densities based on weight (National Farm Animal Care Council, 2025).

As with many farmed animals, the most powerful motivation to improve the health and welfare of turkeys can be traced back to financial losses ensuing from injurious pecking and aggression, footpad dermatitis and leg abnormalities when untenable stocking densities cause physical and psychological harm to the birds. However, increasing consumer outrage at the poor living conditions of these birds is providing further motivation to reconsider their living arrangements.

The negative experiences while confined are compounded by the trauma of catching, crating and transporting turkeys to slaughter. Described as potentially the most stressful events in a turkey's short lifetime, there can also be detrimental economic implications through carcass damage and lost profitability from deaths on arrival (Marchewka et al., 2013). Turkeys may succumb to thermal stress, motion and vibration from the transport vehicle, fasting, lack of water, social disruption, noise and the actual slaughter process. They can arrive with broken bones, dislocations and bruising when their massive size and weight make catching and handling a challenge. Depending on the country, turkeys may be hung upside down by their legs at the slaughterhouse, then moved to an electric bath to be stunned. Pulling their heads through this electrified body of water is intended to render them unconscious before slaughter, but distressed flapping means their wings may hit the water first giving them painful electric shocks before their heads hit the water (Sparks, 2024). The terror these sentient animals must feel is unimaginable, allowing emotional contagion to spread through those listening and awaiting their turn to die.

Despite the neural changes that have been identified in turkeys following domestication, they still retained the capacity to suffer under stressful circumstances. As domestication robbed animals of their right to engage in natural behaviours, consistent research findings suggest that brain size reduced across a range of species. An early study by Ebinger and Röhrs (1995) concluded that male domestic turkeys had around 29 per cent less brain volume and female turkeys around 24 per cent less brain volume compared to their wild relatives. These changes were accompanied by changes in proportions across the main brain subdivisions and regions belonging to the visual system. Wild turkeys have excellent vision to find food, such as insects, nuts, seeds and other food, but importantly, as a primary defence against hunters, predators and other threats. With eyes on the side of their head giving them 360-degree vision with a slight turn of their head, wild turkeys have the ability to see and process visual data very quickly and react appropriately. The domestic turkey's reduction in visual brain structures reflected different survival challenges when confined in overcrowded, stressful intensive farm systems.

FEELINGS AND EMOTIONS

Turkeys have a range of vocalisations that they put to good use when expressing their feelings and communicating with their friends and flock. Gobbling is only one sound in their repertoire: there are clucks, contentment purrs (as Preena demonstrates), yelps to communicate over longer distances, angry or defensive hisses, distressed squeals and turkey poult peeps when they are cold or hungry. They can manipulate these sounds by delivery and intensity to create a complex vocabulary that allows them to express an abundance of different messages and feelings (Talking Birds, 2024). The extent and diversity of their vocalisations become evident among those turkeys living on farmed animal sanctuaries and no longer restricted to expressing fear, distress, frustration and anger.

However, turkeys have more than one way to express their feelings. They can autonomously change their head skin colour in response to their emotions and social interactions, especially when excited or during mating. In Korea and Japan, turkeys have been labelled "seven-faced birds" as the skin on their head changes from red to blue to white, depending on how excited or calm they are. The more intense the colours, especially in males, the more intense their emotions (Oh et al., 2014). Berkeley researchers Oh and colleagues described the bundles of collagen interspersed with a dense array of blood vessels in the turkey's head that gave them this colour-shifting characteristic. The spacing between the collagen fibres changed with the swelling or contraction of blood vessels, depending on how excited, angry or calm the bird was feeling, and the amount of swelling changed the way light waves were scattered, which in turn altered the colours perceived on the turkey's head.

For the turkey friends at Goodheart Farm Animal Sanctuary on the border between Worcestershire and Shropshire in the United Kingdom, there were plenty of opportunities to express their feelings as they settled into a safe and secure environment and formed strong attachments. In the right environment, turkeys are curious, intelligent and quirky, thriving in the company of familiar friends and at complete odds with misconceptions that position them as silly birds reared for celebratory meals.

Living long enough to develop strong friendships is a basic right denied to those turkeys who live and die in industrial farm systems. However, when given the opportunity to live longer lives, this also means that some turkeys will experience the loss of friends, especially when the bones and internal organs of their turkey friends were never meant to support the rapid growth of meat turkeys.

As with any strong bond, there is the risk of losing a beloved friend and experiencing feelings of grief, an emotion that research increasingly confirms is not unique to humans.

Strong bonds—Susan, Colin, Barbara and Mr Biggles, Goodheart Farm Animal Sanctuary, United Kingdom

For five years, Colin, the turkey's multicoloured plumage and fiery personality, captivated all who met him and turned the sanctuary into Colin's personal domain. But Colin had a weakness—his love for Susan. So besotted was he with this gentle lady turkey, he would follow her around and never leave her side, even when her body language suggested she would have preferred a bit of peace now and again.

Susan, a Norfolk Bronze turkey typically raised for meat, was eight months old when she arrived at Goodheart Farm Animal Sanctuary. Originally purchased to fatten up for Christmas dinner, by the time December came around, the owners had learned to care for and respect this sentient animal and could not bear to have her slaughtered. Turkeys are unique, inquisitive and affectionate individuals, traits that are overlooked when confined in overcrowded conditions awaiting slaughter for festive meals. Getting to know Susan saved her life, and the owners brought her to Goodheart's to live out the rest of her life in safety and freedom.

Several months later, Colin arrived at the sanctuary. Colin and Susan formed an enduring bond as they patrolled the sanctuary, snoozed in the woods and relaxed in a shared dust bath. Colin's uncanny ability to find any open door, especially to the feed room, to peck at the shiny wheel caps of visitors' cars or to chase the human team around the yard just for the fun of it cemented his reputation as a cheeky but endearing character. Gentle Susan enjoyed the rewards of his food thefts, while her expressive face and vocalisations were reminiscent of tolerant "embarrassment" at his other antics especially chasing the humans.

Colin's intense emotions regularly saw his snood (above the beak) and wattle (below) move between striking red and blue hues as he raised his feathers to show his size and strength, especially when showing off or causing mischief. Should Susan or the humans ignore this display, his show escalated to shaking tail feathers and demanding gobbles. All this ended in 2023 when Colin passed away after a brief illness.

The loss of Colin's loud, bossy and cheeky personality was felt across the sanctuary. Susan was moved from the main yard to be with Biscuit the goose and her friends in an attempt to alleviate Susan's loneliness and grief. When Barbara, another Norfolk Bronze turkey, arrived at the sanctuary, they formed a friendship that was supportive, but qualitatively different to the one Susan had shared with Colin. Barbara's life was impacted by a pre-existing leg injury and arthritis, and her time at Goodheart's and with Susan was cut short when the pain relief became ineffective. Born to be slaughtered for meat, Barbara died in comfort and safety, so different to the early slaughter of most Norfolk Bronze turkeys.

Figure 9.2 Colin's multicoloured plumage and fiery personality captivated all who met him. (Image supplied by Goodheart Farm Animal Sanctuary and reproduced with their permission.)

Before Barbara's death, the feisty Mr Biggles, a Blue Slate turkey whose loud gobbles and swift pecks to accelerate filling the feeder caught everyone's attention, had joined the sanctuary. Once again, Susan found herself with an eccentric friend, but the two soon formed a bond of friendship and now are often seen sitting side by side on their log perch or exploring in the fields together.

Colin, Susan, Barbara and Mr Biggles moved far beyond their pre-ordained food animal status when relocating to Goodheart Farm Animal Sanctuary. They developed into four different and distinct personalities, formed unique bonds and grieved when death inevitably struck their group. Factory farmed turkeys are deprived of the opportunity to carry out their natural behaviours during their shortened lives. They experience high levels of stress and suffering from confinement and slaughter, ending up as nothing more than a plucked carcass ready to roast for a celebratory meal to give humans a fleeting moment of culinary enjoyment. Quirky duo Susan and Mr Biggles, and Colin and Barbara before them, prove turkeys are so much more than many humans give them credit for.

Unfortunately, domestic turkeys are living examples of the ubiquitous power of human-driven selection as they evolved to provide a consistent, visually appealing, abundant and profitable source of meat for human consumption. Somewhere along the way, their intelligent and playful personalities became invisible within their overcrowded conditions and short lifespans. This was not the case for Penelope from Horse and Farm Animal Sanctuary in Wales when her life trajectory took an unexpected turn.

Figure 9.3 Colin and Susan formed a strong bond and would patrol the sanctuary together. (Image supplied by Goodheart Farm Animal Sanctuary and reproduced with their permission.)

Penelope, a new beginning—Horse and Farm Animal Sanctuary, Wales

In December 2024, a picture of the beautiful Penelope appeared on the Horse and Farm Animal Sanctuary Facebook page with the caption "Penelope will be having her own Christmas dinner, she won't be someone else's…" Penelope's cheeky image travelled from Pembroke in Wales around the world, gathering congratulatory and relieved comments from those who understood the cruel future that had awaited turkeys like Penelope as Christmas approached. These comments were tinged with sadness for the other turkeys who, unlike Penelope, could not escape their human-ordained fate.

Penelope came from an intensive farm setting, one of about 3,000 turkeys being fattened up for Christmas dinner. Understandably timid at first and causing some concern in her first few days at the sanctuary as she sat in a corner with ruffled feathers and no appetite, Penelope was not one to let this opportunity at life pass by. She was after all ambassador for the thousands of turkeys who would not make it past Christmas. Horse and Farm Animal Sanctuary founder, Lisa's love for this affectionate bird was evident as she shared Penelope's story of how every life matters.

Penelope was allowed to be rescued from the overcrowded confines of the turkey farm because she was "poorly" and of no economic value to the producer. Guilt at not being able to rescue more abused animals is an emotion that burdens all those who provide sanctuary to farmed animals. It cannot be eradicated but it can be softened by focusing on one animal—like Penelope—who escapes their fate. As the farmer emerged from the noisy barn with the young turkey swinging upside down by her legs, the bird who would be named Penelope was physically distressed and emotionally shutdown. Transferred to the box that Lisa had prepared, the extent of Penelope's trauma became evident. She had been debeaked, most likely when only one day old, but she also had no snood, which was more unusual.

Snoods are the fleshy flaps of skin that hang over the turkey's beak. Females have smaller snoods than males, and in both sexes, the colour of their snood can be an indicator of their health and emotional state. Females tend to prefer males with longer snoods, possibly because snood length can be predictive of male dominance hierarchies (Buchholz, 1997). Removing snoods does occur in some industrial farm settings to prevent densely packed turkeys from damaging this fleshy vulnerable part of their face, and like debeaking, usually within the first day of life and with no or minimal pain relief (EFSA et al., 2024). The absence of Penelope's snood was a sad reminder of the conditions from which she had escaped and which remained a reality for thousands of other young turkeys.

Slowly, Penelope settled into the sanctuary. Although she still bore the psychological scars of her first four months of life with its horrific overcrowding and rough handling, she came to accept that a new life was possible. The luxury of choice, whether it was deciding where she would make her bed in the deep straw of the turkey house or choosing friends such as resident turkeys Pepper and Pedro, was a new experience for Penelope, but she soon adapted as she learned how to be a turkey. American ecologist and author Carl Safina commented that within animal culture, observation or interaction with the group allows knowledge to be passed from parent to offspring, older to younger, younger to older or between peers of the same generation in the form of an inheritance (Safina, 2020). Being raised in the crowded confines of an intensive farm setting, Penelope did not have the opportunity to receive shared knowledge from her parent, nor could she observe the behaviours of a turkey raised in a natural setting. Penelope's life was scheduled to end at around four months, so time was limited for any social learning beyond the emotional contagion inherent in the stressful environment of intensive farming. Living on a farmed animal sanctuary with friends allowed Penelope to start the process of developing the turkey's natural intelligence, curiosity and affectionate nature and forget the horrors of her previous life.

Very young domestic turkeys have the ability to fly short distances, perch and roost, until their selectively bred rapid weight gain precludes these basic behaviours. In contrast, young turkeys in the wild have been observed spontaneously running, playing among themselves, flapping in dust baths, preening as a group and generally living and learning all the things that young animals need to know in the early stages of their lives. They are also highly social and become distressed when isolated. Having made friends with Pepper and Pedro, Penelope disliked being alone and would cry out whenever she lost contact with her new friends.

Day by day, Penelope's fear passed, and her naturally affectionate nature allowed her to cuddle and nuzzle the neck of the human she now trusted, with whom she was developing the bonds that turkeys are capable of forming if given the chance. Often spotted standing in the sun, watching the wild birds and hopefully finding joy and peace at the sanctuary, Penelope's unique personality started to emerge as she felt more and more confident to explore her new surroundings and experience the world around her.

People also have a choice: they can choose to see turkeys for the complex and unique individuals they are, or they can selectively ignore the right to life for all sentient animals and continue perceiving Penelope and others like her as commodities bred primarily for meat production.

Figure 9.4 Penelope enjoying the sunshine. (Image supplied by Horse and Farm Animal Sanctuary and reproduced with their permission.)

ETHICAL IMPLICATIONS

In line with many other farmed animals, contemporary turkey production systems have been driven by consumer demand and economic practicalities. This has left turkeys at risk of multiple negative welfare outcomes including mobility issues, feather pecking, aggression, wounds, infections and stress. The suffering of turkeys has been acknowledged as one of the most challenging welfare issues in those countries where they are confined and fattened for celebratory meals, including the USA, European Union and Australia, and yet it continues.

In 2004, animal ethicist and philosopher Tom Regan's updated edition of the pivotal book *The Case for Animal Rights* was released. Regan described a series of common grounds for justifying the harms done to farmed animals and thereby avoiding cognitive dissonance. When applied to turkey welfare, these reasons provide a warped rationale for the enduring cruelty inflicted on these sentient animals. According to Regan, some of the most frequently cited reasons included:

- Animal flesh is tasty, and to abstain from eating, it is to forgo certain pleasures of the palate.
- It is our habit, both (perhaps) as individuals and as a culture, to eat meat, and eating it is convenient; to abstain would be to endure the pains of withdrawal and the loss of convenience.
- Some farm animals, most notably chickens and turkeys, are not directly covered by the principles prescribed by the rights view, so farmers are at liberty to treat these animals without regard to these principles (and for similar reasons, consumers are at liberty to eat them) (pp. 333–334).

These baseless rationales may alleviate cognitive dissonance, but are in direct opposition to the rights view, which suggests that all moral agents have certain basic moral rights, and in the case of animals, the fundamental right to be treated with the respect that they are due as a matter of justice. As long as any turkeys remain confined, their fundamental rights as sentient animals are being denied. When the 2012 *Cambridge Declaration on Consciousness* affirmed that birds possess the neuroanatomical, neurochemical and neurophysiological substrates of conscious states and can experience what happens to them, this should have provided impetus to right the wrongs inflicted on turkeys. The stories of Preena, Penelope and Susan, Colin, Barbara and Mr Biggles clearly demonstrate sentience, uniqueness and enjoyment in life, and yet Regan's first two observations continue to provide a major block for many consumers. There remains a reluctance in some people to give up both the taste of turkey flesh and a habit born of long tradition. Neither reason supports the continued manipulation and confinement of these sensitive, affectionate birds.

REFERENCES

AVEC (2024). Turkey management best practices. *European Poultry Meat Sector*. https://avec-poultry.eu/news/turkey-management-guide/

Bartels, T., Stuhrmann, R.A., Krause, E.T. & Schrader, L. (2020). Research note: Injurious pecking in fattening turkeys (Meleagris gallopavo f. dom.)—video analyses of triggering factors and behavioral sequences in small flocks of male turkeys. *Poultry Science, 99*(12), 6326–6331. https://doi.org/10.1016/j.psj.2020.09.016

Bekoff, M. (2024). *The emotional lives of animals: A leading scientist explores animal joy, sorrow, and empathy——and why they matter*. New World Library.

Brant, A.W. (1998) A brief history of the turkey. *World's Poultry Science Journal, 54*(4), 365–373. https://doi.org/10.1079/WPS19980027

Buchholz, R. (1997). Male dominance and variation in fleshy head ornamentation in wild turkeys. *Journal of Avian Biology, 28*(3), 223–230. https://doi.org/10.2307/3676973

Cambridge Declaration on Consciousness. (2012, July 7). Written by Philip Low and edited by Jaak Panksepp, Diana Reiss, David Edelman, Bruno Van Swinderen, Philip Low and Christof Koch. University of Cambridge.

Cambridge Dictionary (2025). *Turkey*. Cambridge University Press and Assessment. https://dictionary.cambridge.org/dictionary/english/turkey

Ebinger, P. & Röhrs, M. (1995). Volumetric analysis of brain structures, especially of the visual system in wild and domestic turkeys (*Meleagris gallopavo*). *Journal für Hirnforschung, 36*(2), 219–28. PMID: 7615926.

EFSA (European Food Safety Authority), Rojo Gimeno, C., Balmos, O.M., Candiani, D.F., Van der Stede, Y., & Fabris, C. (2024). The practice of mutilations in turkeys (*Meleagris gallopavo gallopavo*) – exercise of the EFSA Networks on Animal Welfare. *EFSA supporting publication 2024, 21*(12), EN-9138. 30 pp. https://doi.org/10.2903/sp.efsa.2024.EN-9138

Farm Sanctuary, Inc. (2025). *Five reasons to keep turkeys off your plate this Thanksgiving*. https://www.farmsanctuary.org/news-stories/five-reasons-leave-turkeys-off-plate/

Farm Transparency Project (2020). *Turkeys*. https://www.farmtransparency.org/kb/food/turkeys

Lally, K. (2024, December 25). Real reason why we eat turkey on Christmas Day as tradition explained. *Wales Online* [Cardiff, Wales]. https://www.walesonline.co.uk/news/uk-news/real-reason-eat-turkey-christmas-30624750

Marchewka, J., Watanabe, T.T.N., Ferrante, V., & Estevez, I. (2013). Review of the social and environmental factors affecting the behavior and welfare of turkeys (*Meleagris gallopavo*). *Poultry Science, 92*(6), 1467–1473. https://doi.org/10.3382/ps.2012-02943

Marchewka, J., Vasdal, G., & Moe, R.O. (2019). Identifying welfare issues in turkey hen and tom flocks applying the transect walk method. *Poultry Science, 98*(9), 3391–3399. https://doi.org/10.3382/ps/pez211

Minney, N. (2022). Why do we eat turkey at Christmas? *The Museum of English Rural Life*. https://merl.reading.ac.uk/blog/2022/12/why-do-we-eat-turkey-at-christmas/

National Farm Animal Care Council (2025). *Code of practice for the care and handling of hatching eggs, breeders, chicken and Turkeys*. https://www.nfacc.ca/poultry-code-of-practice#:~:text=This%20Code%20is%20a%20guideline%20for%20the%20care,assess%20animal%20care%20on%20poultry%20farms%20across%20Canada.
National Turkey Federation (2025). *NTF standards of conduct*. https://www.eat-turkey.org/animal-welfare/standards/
Oh, J.W., Chung, W.J., Heo, K., Jin, H.E., Lee, B.Y., Wang, E., Zueger, C., Wong, W., Meyer, J., Kim, C., Lee, S.Y., Kim, W.G., Zemla, M., Auer, M., Hexemer, A., & Lee, S.W. (2014). Biomimetic virus-based colourimetric sensors. *Nature Communications*, *5*, 3043. https://doi.org/10.1038/ncomms4043. PMID: 24448217.
Regan, T. (2004). *The case for animal rights*. University of California Press.
Rink, S. (2023). 9 interesting facts about turkeys. *World Animal Protection*. https://www.worldanimalprotection.ca/blogs/9-interesting-facts-about-turkeys/
Ritchie, H. (2024). How many animals are factory-farmed? *Our World in Data*. https://ourworldindata.org/how-many-animals-are-factory-farmed
Safina, C. (2020). *Becoming wild. How animals learn to be animals*. Oneworld Publications.
Shahbandeh, M. (2025). Global number of turkeys 1990–2023. *Statista*. https://www.statista.com/statistics/1108972/number-of-turkeys-worldwide/
Sparks, M. (2024). What's life like for a factory-farmed turkey? *The Humane League United Kingdom*. https://thehumaneleague.org.uk/article/whats-life-like-for-a-factory-farmed-turkey
Talking Birds (2024). *The ultimate guide to Turkeys: Everything you need to know!* https://www.talking-birds.co.uk/the-ultimate-guide-to-turkeys-everything-you-need-to-know/#:~:text=Key%20Turkey%20Sounds%3A%201%20Gobble%3A%20The%20most%20iconic,a%20series%20of%20single-note%20vocalizations.%20...%20More%20items
Walk, C.L., Alleno, C., Bouvet, R., Thoby, J.-M., Eising, I., & Segobola, P. (2023). Dietary muramidase improved growth performance, feed efficiency, breast meat yield, and welfare of turkeys from hatch to market. *Poultry Science*, *102*(7), 102716. ISSN 0032-5791. https://doi.org/10.1016/j.psj.2023.102716

10

Geese: Intelligent, emotional and charming

INTRODUCTION

Many fables immortalised through books and film feature animals behaving and speaking as humans in order to highlight human foolishness or weaknesses. *The Goose that Laid the Golden Egg*, commonly ascribed to ancient Greek storyteller Aesop (620 and 564 BCE), is one such story that provides a reminder of what can happen when human greed overcomes compassion (Aesop & Perry, 1952). The untimely end of the goose who laid the golden egg highlights humanity's relentless ability to destroy the natural world as they constantly seek more at any cost.

The protagonist, a Countryman, owned a wonderful goose who laid a big egg made of pure, shiny, solid gold every morning. Slowly, the Countryman became rich from the daily offering, but it was not fast enough. He wanted instantaneous wealth by getting all the eggs at once that he believed must be hidden away within the goose. After pondering possible options, the Countryman decided to kill the goose and cut her open, only to find no eggs inside her and no more daily golden eggs. The Countryman was not the first human who destroyed something important in the natural world for perceived short-term personal gain.

While the goose's ability to lay golden eggs is simply a fable, humans have used waterfowl—of which geese are members—for thousands of years in multiple ways. Their flesh and eggs are eaten, their unnaturally fattened liver is sold in the form of foie gras at exorbitant prices, their fat is used for lighting and heating and their feathers and down are plucked from their often-live bodies to provide warmth and fillers in bedding and clothing. In the Jewish diet, goose fat can play an important role as a pig-fat substitute, while it was also used as a medicine in ancient civilisations (Kozák, 2021). In fact, the goose was turned into a "multi-purpose winged animal" (p. 403). When not being eaten or plucked, their natural vigilance and tendency towards fierce protectiveness saw them turned into watchdogs as far back as 390 BCE, right through to forming the "Scotch

DOI: 10.1201/9781003518853-11

Watch" that guarded the Scottish Dumbarton grain whisky distillery between 1959 and 2002 (Hemetsberger et al., 2013).

Under the right conditions, domestic geese can live for 25–30 years, but even greater longevity has been reported on occasion. In England, a gander died at almost 50 years of age, while an unconfirmed report of a goose passed down through three family generations in Massachusetts USA claimed the age of 101 years (Kozák, 2021). In commercial breeding situations, geese are kept for considerably less time, being killed at an average five years of age.

Although lower in number compared to other commercial poultry birds, domestic geese can be more adaptable than chickens to a range of environmental conditions. Located in China, the world's leading goose production area, as well as Myanmar, the Russian Federation, Ukraine, Madagascar, Egypt, and Türkiye, geese comprise less than 1.3 per cent of the global poultry population (Yan et al., 2025). Following a similar trajectory to other intensively farmed animals, humans identified traits most beneficial to their needs and selectively bred for those traits to maximise the commercial benefit of keeping geese.

SOME BACKGROUND HISTORY

Only two species of the numerous wild true geese have been domesticated: the Greylag goose (*Anser anser*) and the Chinese Swan goose, recognised as the ancestor of the Knobbed goose (*Anser cygnoides f. domestica*) (Kozák, 2021). European geese derive from the Greylag goose and range in colour from the original wild grey shades to white, while South-east Asian geese derive from the Chinese Swan goose. The Knobbed goose usually has a large knob at the base of their bill and can vary from mostly grey to pure white. Other species were tried but rejected when it became clear that only the Greylag and Chinese Swan geese had traits that matched human needs for domestication.

More than 20,000 years ago, drawings of Greylag geese started appearing in numerous civilisations including South-West Asia, China, South-East Asia, North Siberia, Mesopotamia, Asia Minor, Egypt, ancient Greeks and the Romans, suggesting that geese may have been one of the earliest birds to be selected for domestication (Kozák, 2021). Domesticated geese are now found in every part of the world, with some crossing of the two breeds. Charles Darwin commented that compared to other farmed species, few had varied so little as the goose after such a long period of domestication—but that would change in more recent years.

Excessive breeding for selected traits, some of which had negative implications for the goose's wellbeing, meant that current domestic varieties vary in appearance and features. Higher meat yield and faster growth, high fat storing ability, excess egg production and ease of feather harvesting all increased the economic viability of farming these majestic birds, thus facilitating their relocation to all continents of the world. Breeding for increased weight made getting off the ground and flying for any distance difficult, forcing them to walk more on load-bearing bones (femur, tibiotarsus, tarsometatarsus), which by necessity became stronger. In the Greylag goose, length measurements of these bones remained almost identical to those of their wild relatives, with only the breadth and heaviness changing, but wing bones became lighter, and musculature decreased by about 30 per cent (Kozák, 2021).

Despite human manipulations, domestic goose breeds have not adapted well to large flock keeping. Their sensitivity to the negative effects of stock density can impact feather quality and overall wellbeing, with flow-on economic implications for the producer. Another barrier to the large-scale production approach taken with most other commercial poultry can be the poor reproductive performance of male breeder stock and the resultant lower fertility and hatchability rate (Yan et al., 2025). In fact, geese have the lowest reproductive efficiency of all domesticated poultry species. In the wild, geese form a life-long pair bond, with annual sexual activity repeated for a short time in Spring so that offspring can be born in a season most favourable to their survival. Seasonal reproduction is therefore an evolutionary strategy that can result in economic losses in the poultry industry (Liu et al., 2023). Unable to leave well alone, human manipulation has seen some geese reach sexual maturity as early as seven or eight months, thus shifting the monogamous pair-bond of wild geese towards polygamy and egg production starting at the young age of eight or nine months (Kozák, 2021).

GEESE IN CONTEMPORARY SOCIETY

Modern-day domestic geese need to grow very big, very quickly. In fact, they have been selectively bred to have one of the fastest growth rates in domesticated poultry species, a physiological manipulation assisted by early fattening techniques (Kozák, 2021). They are grazing birds whose dilatable gullet allows them to take in large quantities of food in a short time. Unfortunately, this ability also allows young geese to be overfed on green forage to dilate their digestive tract in preparation for even higher feed intake during fattening times. It can also be a precursor to the abusive force feeding required to create surplus fat deposits in the liver to make the luxury food item, foie gras. Even when spared the intensive farming methods inflicted on other poultry, geese can still be abused through the relentless commodification of their body parts.

Foie gras remains a delicacy in many parts of the world despite public outcry at the horrendous abuse inflicted prior to harvesting the grossly fattened livers. Ducks are also subjected to a similar forced feeding regime, but goose fatty liver is considered of better quality for making foie gras. This belief dates back more than 2,500 years when cramming of geese emerged in ancient Egypt and was subsequently adopted by Greeks and Romans (Guèmenè & Guy, 2004). Egyptians no longer force-feed geese, but it remains a tradition in some parts of Europe where the majority of fatty liver production is concentrated. Following World War 2, French foie gras production became industrialised to allow year-round output, lower costs and encourage consumer demand.

Force-feeding is as horrible as it sounds. A pipe is pushed down a captive goose's throat several times a day to pump them full of food and make their liver swell to unnatural and unsustainable sizes. As more and more countries banned this archaic process, there remained pockets of resistance where production was legal despite the obvious disregard of animal welfare standards. To be marketed as foie gras, the goose's liver must reach a minimum weight of 400 grams (14 ounces), but often more, which can be a tenfold increase in size and with a fat content of over 50 per cent (Guèmenè & Guy. 2004). To put this in context, a median liver weight in humans is 1,400 grams (3 pounds) for women and 1,800 grams (4 pounds) for men. If a human liver was subjected to the same unnatural cramming as geese, people would be carrying round weights of 14–18 kilograms (30–40 pounds) in their abdomens.

Fatty liver production cruelly distorts a goose's natural development to overeat and store excess energy in their livers prior to the long migratory flight. But these geese are going nowhere, instead trapped in cramped pens and unable to move as their livers swelled to unnatural sizes. In his updated book *On Food and Cooking*, well-known food scientist and writer Harold McGee (2004) described foie gras as "It's a kind of living pâté, ingeniously prepared in the growing bird before it's slaughtered" (p. 271). Ingenious is hardly an appropriate word to describe the process of shoving increasing amounts of grain down the goose's throat through a specialised tube or pipe. This process, called "gavaging" from the French word "gavage", to gorge, lasts for 12–21 days of discomfort and misery for the geese, some of whom succumb to internal organ failure and death as they approach their maximum liver weight (Kolpas, 2021).

Fatty livers are not the only part of a goose's body harvested for human use. Goose feathers have been prized by humans for over 2,000 years, with the Romans learning from the German and Celtic tribes about the versatility of these birds' feathers and down (Kozák, 2021). From that point on, goose feathers and down, the soft layer of feathers closest to their skin, became valued commodities for their lightness, long-lasting durability (around 50 years) and unique insulation properties in bed linen, winter garments including ski clothes and sleeping bags. Goose feathers can be acquired from dead birds at the slaughterhouse or gathered from live birds at moulting time, but high demand has seen some geese plucked while still alive causing unimaginable pain and suffering. Unfortunately, feathers collected from the belly, chest, back flanks and lower part of the neck of live birds are considered of better quality, thus driving demand. These feathers can only be gathered manually to avoid injury from mechanical removal, an ironic observation considering the suffering inflicted on these birds as feathers are ripped from their bodies.

Some countries, including the European Union, placed a ban on the plucking of feathers and down from live birds, but graphic images released by organisations such as People for the Ethical Treatment of Animals (PETA) and Four Paws indicates some farmed geese continue to have their feathers "violently pulled out of their skin every 6–8 weeks"—often illegally—in countries with or without bans (Four Paws Australia, 2022). This may start in goslings as young as ten weeks and also provides additional income for the foie gras farmers who have been known to sell birds' feathers before they are slaughtered.

PETA UK (2025) provide an emotive description of what it means to endure live plucking:

> Imagine being lifted up by your neck and having your hair torn from your head without painkillers [...]. The struggling birds are roughly pinned down and have their feathers yanked out so hard that their skin is often torn open, leaving gaping bloody wounds.

Irrespective of where the plucking occurs, other countries become duplicitous when importing products originating from areas that allow birds to be treated in this way. Being removed from the source of the cruelty may be sufficient to allay feelings of guilt and cognitive dissonance, but as long as there is a market for goose feathers and down, these birds remain at risk of welfare breaches.

The commodification of domestic geese goes beyond force feeding and feather plucking to an extension of the egg laying period through manipulation of light conditions (Kozák, 2021). Under natural light conditions, breeder geese lay eggs for 20–24 weeks, usually every second day. By introducing a constant nine to eleven-hour lighting system, the laying period can be extended by 30–40 days, increasing egg yield. A second laying cycle can also be induced in Autumn when geese should be in a state of rest under natural conditions, increasing the laying period and stress on the goose's body. As with chickens who are relentlessly manipulated to lay an unnatural number of eggs, this cycle will inevitably impact the health and wellbeing of the layer goose.

Even when geese go off to slaughter, human interests may dictate methods to preserve the best quality of meat and liver (Fernandez et al., 2010). As often occurs in farmed animal slaughter, research may be driven by quality of the product as much as animal welfare. A report on the various stunning methods used for geese, including water-bath electrical stunning, controlled atmosphere stunning and mechanical stunning, focused on the potential effects on the quality of the carcass, meat and fatty liver (in foie gras geese), highlighting that the final stages of a goose's life must still conform to human taste and the visual appeal of the end-product.

FEELINGS AND EMOTIONS

Brucey the Goose, whose story appears later in this chapter, is a charming protective gander to his gaggle of females. Brucey is now living his best life at Goodheart Farm Animal Sanctuary in the United Kingdom where each evening he enjoys a bath in the paddling pool, a filling meal and time to sit back and relax while he and his friends preen their beautiful feathers, which will never be plucked from their living or dead bodies.

Like all birds, Brucey lacks a clearly defined neocortex—that part of a mammal's brain that deals with working memory, planning and problem solving. This has seen birds denigrated with labels such as "bird brain" and deemed incapable of complex cognition due to their mainly striatal forebrain. Because there was a belief that the avian telencephalon (cerebrum) was composed of mostly basal ganglia, the perception lingered that all their actions were instinctive. However, microcircuits in these bird brains are increasingly being investigated as analogous to the mammalian neocortex, with some researchers linking these microcircuits to conscious thought. Some areas once labelled as primitive avian pallial areas—that is, the hyperpallium, the nidopallium and the arcopallium—are processing regions with functions similar to the mammalian cortex and mammalian pallial regions (Scheiber et al., 2013). That means even though bird and mammal brains evolved differently, they experience and respond to their physical and social environment in similar ways.

Researchers Andrew Iwaniuk and Peter Hurd (2005) suggested that avian species could be sorted into defined cerebrotypes—that is, by types of brain composition shared between groups and reflecting specific ecological niches or evolutionary relationships. Waterfowl were linked to Cluster D cerebrotypes, which also comprised parrots and crows with their relatively large telencephalic volumes. While considerable research supports the potential for complex cognitive abilities and social interactions among parrots and crows, waterfowl—including geese—have received less research focus. However, based on brain composition, it is possible that geese also have capacity for a range of complex thought processes denied to them when valued only for the products their bodies provide.

A common accompaniment to domestication is reduced brain size, and domestic geese are no exception. Domestic geese brains have reduced by 15–16

Figure 10.1 Access to water is essential for goose wellbeing, as this Greylag goose paddling on a pond demonstrates. (Image supplied by Rebecca Hutton and reproduced with permission.)

per cent to represent around 0.15 per cent of their body weight (Ebinger & Löhmer, 1987), with the largest reduction in size being the hippocampus, located in the avian pallium (Scheiber et al., 2013). Spatial learning and memory are associated with the hippocampus, and both these aspects may be impaired by the removal of geese away from their natural environments. Exclusion of any animal from their natural habitat necessarily results in altered behavioural instincts related to food seeking, defence against predators and survival in general. When unable to gather at the local pond, paddle around and stick their head under water to clean their nasal passages and eyes, confined geese may experience a deterioration in faculties and wellbeing. This is compounded by their inability to fly long distances weighed down by the heavy bones required to support their increased body weight. Unfortunately, domestic geese will never experience the freedom of flight enjoyed by their wild relatives.

Many urban dwellers will never meet a goose in their natural environment, and so their perception of the goose's personality and cognitive skills may be based on stereotypes. These stereotypes remove an animal's individuality and instead rely on inferred characteristics attributed to the whole group and used as a shortcut to allow rapid responses or opinions. Human racism can be fuelled by stereotypes, as can the demarcation between attributions of unique personality traits to familiar companion animals and the inability to differentiate unfamiliar farmed or wild animals. Geese have a reputation for aggressiveness, often preceded by a warning hiss. Beyond this aggressive stereotype, there is a naturally wary animal protecting their nest, eggs, goslings or partner from perceived danger and anyone or anything that ignores their preliminary "back off" hiss.

Austrian zoologist, ethologist and ornithologist Konrad Lorenz's seminal work among Greylag geese provided insight into geese as individuals, not stereotypes. Images of Lorenz being followed by young goslings are almost mandatory inclusions in both human psychology and animal behaviour texts as his work on imprinting earned him a share of the 1973 Nobel Prize in Physiology or Medicine. Lorenz claimed that all the geese differed in their behavioural tendencies and that every goose had a distinct personality, a finding that has subsequently found empirical support from a phenomenological, developmental and physiological perspective over the years (Kralj-Fišer et al., 2013).

Wild geese form couples with a tight life-long bond remaining until the death of one of the partners (Zaynagutdinova et al., 2020). Whether feeding, resting, socially interacting, guarding or migrating, the male and female are together every day, maintaining a proximity of approximately two metres (6.5 feet). To constantly protect and keep track of each other, they employ a range of vocalisations, ranging from hissing to ward off animal predators and annoying humans, to the cheerful honking between members of the same gaggle as they greet their friends or mate.

For wild geese, parental care is also a long-lasting bond, with hatchlings imprinting on their parents in the immediate post-hatch period. If the parents are not there, the goslings will imprint on the first creature they see, as Konrad Lorenz so eloquently demonstrated in his Nobel Prize winning investigation among Greylag geese (Hemetsberger et al., 2013). Lorenz and his colleagues conducted a longitudinal study of the social developments among long-lived greylag geese from 1950 onwards, eventually establishing a flock at the Konrad Lorenz Research Station in Austria in 1973.

During these observations, it became evident that while Greylag geese typically form heterosexual, monogamous pair bonds, variations are also seen. Male–male pairs and trios (female–male–male or female–female–male) could be formed flexibly over the course of individual life histories (Weiß, 2013). Males tended to be more aggressive and high-ranking than females, and social status was positively correlated with aggressive motivation and success—in many ways reminiscent of human social norms. Wild geese follow an innate nesting instinct and parental care, which cannot be duplicated in large-scale animal agriculture systems where artificial hatching of eggs and commercial rearing of day-old goslings have negated the function of parental care and attachment (Kozák, 2021).

Removing geese from their natural social systems and environment deprives them of social support, leaving them at risk of activating the physiological stress response. A goose's personality, life history, social environment and parental genetic and non-genetic influences all contribute to successful pair bonding, reproduction, gosling survival and overall wellbeing. Levels of immune-reactive corticosterone metabolites can increase without the presence of a supporter—even if there is no active involvement—highlighting the welfare implications of large-scale production systems for geese (Scheiber, 2013).

Brucey the Goose is a testament to the power of a supportive presence. Brucey's life entered a new and exciting chapter when he joined the gaggle of geese at Goodheart Farm Animal Sanctuary.

Brucey and friends—Goodheart Farm Animal Sanctuary

In June 2022, Goodheart Farm Animal Sanctuary, a 37-hectare (92 acre) site set in the rolling countryside on the border between Worcestershire and Shropshire in the United Kingdom, welcomed a new resident, Brucey the goose. Based on the belief that all animals are individual sentient beings capable of feeling both positive and negative emotions, Goodheart founders David Walker and Dwynwen Jones set out to action this belief among "the underdogs of the animal world—farmed animals" (Goodheart Animal Sanctuaries, n.d.). The Nickless, in Worcestershire, surrounded by gentle hills, wooded areas and large pastures, provided the perfect setting for Goodheart Farm Animal Sanctuary in 2017. The sanctuary would become home to over 300 rescued animals including pigs, goats, cows, sheep, geese and chickens, confirming the founders' belief in the need to protect the underdogs.

But back to the story of Brucey the goose. Brucey and his mate had been dumped on a mere (shallow pond, lake or wetland) that was subject to blue green algae. Toxic to humans and animals if ingested, Brucey was the lone survivor after his mate disappeared. An urgent call from a nearby resident would change Brucey's life. "He's alone and very miserable", the resident explained, having desperately sought to save him from illness and starvation before it was too late.

Goodheart heralded a new life for the grieving Brucey. This charming elderly white goose with his unique tuft of feathers on the top of his head quickly became a charismatic guardian, protector and support for his gaggle of female geese, including Biscuit, Jemima, Puddles and Scarlet. From a goose whose sad emotions were etched in his posture and vocalisations, Brucey's twilight years were suddenly converted to comfort and safety. Goodheart is home to two flocks of geese. The wild geese keep to themselves on their own little island in the main pool, while Brucey and the tame flock patrol the yard and the smaller garden pool. Every morning, Brucey and his friends leave the night-time safety of their yard and waddle down to their pond for a day of bobbing, bottoms up and grazing in the nearby fields of the sanctuary. Every evening as the tame geese return to their evening roost to keep them safe from wandering foxes, Brucey can be seen enjoying a quick pre-bed bath in the little paddling pool, a delicious meal of corn and time spent sitting back and preening his beautiful white feathers before turning in for the night.

All those at Goodheart Farm Animal Sanctuary have no doubts that Brucey and the other 300 animals are unique individuals who have capacity to feel pain and sadness, joy and comfort. Not being able to articulate this in human words does not mean these feelings do not exist. They are clearly visible to anyone who chooses to look. Brucey's rescuer saw a sad individual who had lost his mate. Brucey did not say this, but his body language was evidence enough to trigger a kind human's feelings of empathy. Brucey's story is not unique at Goodheart, where each animal arrives at different stages of life and past experiences. Each is treated as an individual with specific needs and care plans in a safe haven where animals can live in a natural or near-natural environment, free from exploitation.

Figure 10.2 Brucey and his flock grazing in the nearby fields of the sanctuary. (Image supplied by Goodheart Farm Animal Sanctuary and reproduced with their permission.)

ETHICAL IMPLICATIONS

While geese may have escaped the mass exploitation commonly associated with layer chickens and broilers, they are not immune to humanity's whims and desires. Wild geese stretch their wings and migrate, often causing havoc during breeding season when they fiercely defend their breeding grounds and attack humans. This grumpy reputation has followed domestic geese who will never know the fierce independence of their predecessors. Instead, they are grounded when their bodies become too large to fly, bloated from the fattening of their livers or stripped naked for the softness of their feathers and down.

Beneath the stereotype of grumpy aggressive birds who happen to have the best livers for foie gras and the best feathers and down to keep humans warm, they are intelligent, emotional and charming individuals as Brucey's story illustrates. After becoming one of the earliest poultry birds to be domesticated by humans, geese have more than paid their due to humanity. An increasing number of countries and regions have banned force feeding and live plucking, but cognitive dissonance provides a loophole by removing personal responsibility for those consumers who benefit from imported products. The rationalisation that this does not happen in my country is flawed. Irrespective of where geese are crammed or plucked, it becomes a global responsibility to recognise the inhumanity of these procedures and thus eliminate demand. Brucey and others like him matter in a world that remains dominated by human perceived need and greed.

REFERENCES

Aesop, & Perry, B. E. (Ben E. (1952)). *Aesopica : a series of texts relating to Aesop or ascribed to him or closely connected with the literary tradition that bears his name; collected and critically edited, in part translated from oriental languages, with a commentary and historical essay*. University of Illinois Press.

Ebinger, P. & Löhmer, R. (1987). A volumetric comparison of brains between greyland geese (*Anser anser* L.) and domestic geese. *Journal für Hirnforschung*, 28, 291–299.

Fernandez, X., Lahirigoyen, E., Bouillier-Oudot, M., Vitezica, Z., & Auvergne, A. (2010). The effects of stunning methods on product qualities in force-fed ducks and geese. 2. Fatty liver quality. *Animal*, 4(1), 139–146. https://doi.org/10.1017/S1751731109990851

Four Paws Australia (2022). *New shocking footage reveals live plucking of geese for common fashion products continues despite calls for change*. https://www.four-paws.org.au/our-stories/press-releases/2022/shocking-footage-live-plucking-of-geese-for-fashion

Goodheart Animal Sanctuaries (n.d.). About us. https://www.goodheartanimalsanctuaries.com/about-us/

Guèmenè, D. & Guy, G. (2004). The past, present and future of force-feeding foie gras production. *World's Poultry Science Journal, 60*, 210–222. https://doi.org/10.1079/wps200414

Hemetsberger, J., Weiß, B.M., & Scheiber, I. (2013). Greylag geese: From general principles to the Konrad Lorenz flock. In I. Scheiber, B. Weiß, J. Hemetsberger, & K. Kotrschal (Eds.), *The social life of Greylag geese: Patterns, mechanisms and evolutionary function in an avian model system* (pp. 3–25). Cambridge University Press.

Iwaniuk, A.N. & Hurd, P.L. (2005). The evolution of cerebrotypes in birds. *Brain, Behavior and Evolution*, 65, 215–230. https://doi.org/10.1159/000084313

Kolpas, N. (2021). *Foie Gras. A global history*. Reaktion Books.

Kozák, J. (2021) Goose production and goose products. *World's Poultry Science Journal*, 77(2), 403–414. https://doi.org/10.1080/00439339.2021.1885002

Kralj-Fišer, S., Daisley, J.N., & Kotrschal, K. (2013). Individuals matter: Personality. In I.B.R. Scheiber, B.M. Weiß, J. Hemetsberger, & K. Kotrschal (Eds.), *The social life of Greylag Geese: Patterns, mechanisms and evolutionary function in an Avian model system* (pp. 45–64). Cambridge University Press.

Liu, J., Xu, Y., Wang, Y., Zhang, J., Fu, Y., Liufu, S., Jiang, D., Pan, J., Ouyang, H., Huang, Y., & Tian, Y. (2023). The DNA methylation status of the serotonin metabolic pathway associated with reproductive inactivation induced by long-light exposure in Magang geese. *BMC Genomics*, 24, 355. https://doi.org/10.1186/s12864-023-09342-0

McGee, H. (2004). *On food and cooking. The science and lore of the kitchen.* Scribner.

PETA UK (2025). *Down: Ducks and geese pay the price.* https://www.peta.org.uk/issues/animals-not-wear/down/

Scheiber, I. B. R. (2013). 'Tend and befriend': The importance of social allies in coping with social stress. In I.B.R. Scheiber, B.M. Weiß, J. Hemetsberger, & K. Kotrschal (Eds.), *The social life of Greylag Geese: Patterns, mechanisms and evolutionary function in an Avian model system* (pp. 156–171). Cambridge University Press.

Scheiber, I.B.R., Kotrschal, K., & Weiß, B.M. (2013). The greylag goose as a model for vertebrate social complexity. In I.B.R. Scheiber, B.M. Weiß, J. Hemetsberger, & K. Kotrschal (Eds.), *The social life of Greylag Geese: Patterns, mechanisms and evolutionary function in an Avian model system* (pp. 191–201). Cambridge University Press.

Weiß, B. M. (2013). Alternative social and reproductive strategies. In I. B. R. Scheiber, B. M. Weiß, J. Hemetsberger, & K. Kotrschal (Eds.), *The social life of Greylag Geese: Patterns, mechanisms and evolutionary function in an Avian model system* (pp. 88–104). Cambridge University Press.

Yan, L., Liu, J., Chen, R., Lei, M., Guo, B., Chen, Z., Dai, Z., & Zhu, H. (2025). Reproductive characteristics and methods to improve reproductive performance in goose production: A systematic review. *Poultry Science, 104*(6):105099. https://doi.org/10.1016/j.psj.2025.105099

Zaynagutdinova, E., Karenina, K., & Giljov, A. (2020). Lateralization in monogamous pairs: Wild geese prefer to keep their partner in the left hemifield except when disturbed. *Current Zoology*, 67(4), 419–429. https://doi.org/10.1093/cz/zoaa074

11

Ducks: Water-loving and social individuals

INTRODUCTION

According to People for the Ethical Treatment of Animals (PETA) Australia (2024), about 300,000 ducks are legally "blasted out of the sky" for sport every year during the annual duck hunting season in Victoria, Australia. Ironically, those killed outright are the lucky ones. The unlucky ones, often shot by amateur hunters, are left wounded to slowly die in agony. Some estimates suggest one in every four birds who are injured will suffer for hours, days or even weeks before dying (Animals Australia, 2025).

Meanwhile, Australia's duck meat industry has been steadily growing, with eight million ducks sent to slaughter annually from commercial operations. These intensive indoor housing systems can hold tens of thousands of ducks with no access to open water sources and unable to perform their natural behaviours (RSPCA, 2024). Ducks are curious, communicative individuals who feel pain and fear like all animals, and yet in Australia and many other countries, their life choices and longevity are limited.

Global domestic duck populations have steadily increased over several decades, rising from 193 million in 1961 to reach 1.1 billion ducks raised for meat or eggs, and stock ducks for breeding by the twenty-first century (Worldostats, 2025). Contributing factors include advances in breeding techniques and expansion of duck farming in Asia, particularly China, to meet local and global demand for duck products. China maintains the largest duck production and consumption in the world, with around 691 million ducks (Yang et al., 2020), well ahead of the next highest locations, Vietnam at around 84 million, Bangladesh at 64 million and Indonesia at just over 58 million. However, when looked at in terms of duck population per capita, Vietnam has 837 ducks per 1,000 people, compared to China's 488 ducks per 1,000 people, reflecting Vietnam's reliance on ducks as an essential food source. As the practice of cage-rearing laying ducks increased in some areas, the negative impact on these sentient birds became significant (Sun et al., 2024).

DOI: 10.1201/9781003518853-12

The growth in Asian populations mirrors trends in the global poultry industry and increased consumer demand in western nations for a range of protein sources. By 2019, over 27 million ducks in USA and almost 180 million ducks in the European Union were being slaughtered each year to meet this demand (Makagon & Riber, 2022). As their popularity increases, so too does the potential for life to get harder for ducks around the world.

SOME BACKGROUND HISTORY

A combination of existing archaeological evidence and increasing genome data has resolved some questions about the domestication of ducks but left others unanswered. There are two species of domestic duck: the common domestic duck (*Anas platyrhynchos domesticus*) stemming from the wild mallard (*Anas platyrhynchos*) in Southeast Asia, and the Muscovy duck (*Cairina moschata*) from South America.

Around 2,200–2,500 years ago domestic ducks started diverging from the wild mallard ducks in Southeast Asia, especially the southern part of China (Jiang et al., 2021). Whole genome sequencing on a group of domestic and mallard ducks in China suggested that the domestic ducks differed in the expression of genes related to plumage, fat metabolism and neurological function. Mallards are migratory birds, but as wing length in domestic mallard ducks reduced, so too did their capacity to fly and their freedom to migrate (Herrera-Castillo et al., 2022). Several genes related to developing brain and nervous system, feather colouring and meat production during domestication produced ducks who were more docile and less vigilant, with one breed—the Pekin duck—a valuable addition to many communities for their white feathers, large body size, plump breast muscles, fat deposits and high egg yields. Domestic ducks in general gained a reputation as one of the most successfully domesticated birds and a major source of meat, eggs and down feathers for human use.

Meanwhile, the Muscovy ducks who resided in the swampy habitats and tree-lined riverbanks of South America, between southern Mexico, Peru and northern Argentina, were domesticated by the native Colombians and Peruvians (EFSA et al., 2023). Spanish and Portuguese explorers introduced these ducks to Europe in the sixteenth century, as they simultaneously spread to Central America, the southern part of North America and around the world. Two origins of the name Muscovy duck have been suggested: a version of "musk" duck based on its musky smell, or of "Miskito" duck after the Mosquito coast of South America where the Spanish first met these popular ducks. Domesticated Muscovy ducks retained some ability to fly, but never fully lost their anti-predator responses such as freezing, alarm calling, running from danger and vigorous struggling if caught, handled or force fed. They will fight using their claws, wings and bills, especially if chasing off intruders.

During domestication, a species is selectively manipulated to suit human requirements, and the duck was no exception. Domestic ducks no longer needed to evade the predators of their wild relatives, and so retained various plumage colours that in the wild would have made them vulnerable. The prized white plumage of the Pekin duck literally made them a "sitting duck" target and highlighted the death sentence imposed on any domestic duck discarded in parklands and lakes when people had no further use for them.

Neither Neil Diamond nor Garry were stark white Pekin ducks, but they were not immune to the multiple dangers of abandonment. Once healing within the safety of Edgar's Mission in Victoria, Australia, the boys established a mutually supportive bond and demonstrated a forgiving spirit that many humans would struggle to emulate.

Neil Diamond and Garry, best friends—Edgar's Mission, Australia

Fishhooks are a danger to all animals, not just fish, so when a call came in about an abandoned domestic duck with a fishing hook embedded in his mouth, the Edgar's Mission team swung into action. Locating the duck on a picturesque lake in the Melbourne outer suburb of Caroline Springs was the easy part. The mission to rescue this clever brown duck would eventually draw in three rescue groups, 32 people, four synchronised attempts and a dedicated veterinarian before success. An emergency late-night dash to the vet revealed the barbed hook, with its long trailing length of fishing line, was lodged in the duck's cheek, not his throat, and quickly removed. Once free of this harsh reminder of human recklessness, the newly named Neil Diamond was soon quacking cheerfully and waddling confidently around the people that he had previously eluded so easily.

Meanwhile, back at Edgar's Mission, a duck named Garry was unaware that his lonely, searching quacks would soon be alleviated. Domesticated ducks like Gary and Neil Diamond are deeply social beings. In the words of Pam Ahern, founder of Edgar's Mission, "Their joy lies in togetherness—in the rhythm of shared space, mirrored movement, and the harmony of returned quacks. They bond through body language, touch and gentle rituals—like grooming, and splashing about a pond".

Neil Diamond was introduced to Garry, and a friendship quickly formed as they shared their healing journey within the safety of the sanctuary. Garry's iridescent black plumage streaked with shimmers of emerald and violet identified him as a non-native Cayuga duck. Cayuga ducks, originally from the USA, are known for their calm and docile temperament making them popular as pets, show birds or starter birds for novices. Garry's gentle nature and lack of survival instincts meant he had not fared well on his own, leaving him open to the predator attack that almost claimed his life.

The story of Neil Diamond and Garry contrasts two competing aspects of human society: the humans who abandoned them as "worthless" objects, and the humans who saw them as unique individuals worthy of compassion and care. Many domesticated animals are forced to suffer and die in silence, partly because to recognise their suffering is not compatible with an anthropocentric perception of superiority, succinctly summed up by philosopher Melanie Challenger (2021): "Little wonder then that we have spent much of history asserting that human experience has a meaning and value that is lacking in the rigid lives of other animals" (p. 3). Imbuing Neil Diamond, Garry and the millions of ducks abandoned, hunted or slaughtered each year with potential for a meaningful life is to invite cognitive dissonance on an unbearable scale for some people.

However, watching Neil Diamond and Garry's joy in life, forgiveness and deep friendship makes it impossible to view these ducks as anything other than sentient caring beings worthy of living in peace and safety. The team at Edgar's Mission who knows these boys so well has provided insight into their new lives with people who care:

> For now, Garry is still recovering from his ordeal. His back wounds are healing, and his spirit is intact—though a tad dented. But already, his gentle personality is quacking its way to the surface (albeit softly, as only the female ducks truly quack). He adores his pond, gliding across it with regal aplomb and dipping his beak into the water with quiet contentment. As he stands and proudly flaps his wings, we honour his wishes and promise him vaster shores await. In the meantime, he delights us all, carefully picking his favourite greens from his bowl—a simple act, yet one that speaks volumes about the joy of second chances. Together, at sanctuary, they [Neil Diamond and Garry] have found friendship, freedom and fortitude. As they shuffle along, side by side, their chatter is incessant, like long-lost friends reunited. And watching them, we cannot help but think they are quacking: 'Thank the heavens the good guys came to save us'.
>
> *[Reproduced with permission from Edgar's Mission, Australia]*

Figure 11.1 Neil Diamond (left) and Garry stand together as they share joy in life and deep friendship. (Image supplied by Edgar's Mission and reproduced with their permission.)

DUCKS IN CONTEMPORARY SOCIETY

For thousands of years, ducks in China were raised in traditional waters (semi-grazing) and rice paddies, making efficient use of agricultural lands. As growing numbers threatened local communities with increased pollution in rivers and water sources and the spread of infectious diseases, large-scale high-density indoor feeding systems were established to meet growing demand and save costs. This shift from traditional systems to confinement in many countries raised significant welfare concerns for these independent water-loving birds (Yang et al., 2020).

High density may bring potential economic benefits, but abnormal behaviours such as stress convulsions and pecking addiction are a cruel side effect to limited moving space and distance between ducks. Increased blood corticosterone levels and reduced antioxidant activity among overcrowded and stressed ducks impact all facets of their life, from production performance and product quality to disease prevention and control (Yang et al., 2020). Like chickens, ducks were selectively bred to improve food conversion rates and reach slaughter weight unnaturally fast. That meant inadequate skeletal development was unable to keep up with muscle growth, resulting in gait abnormalities, lameness, painful fractures, splayed legs and suffering.

Ducks farmed for meat are typically ready for slaughter at around 3.5–4.5 kilograms (7.7–9.9 pounds) in weight. Selectively bred Pekin ducks reach the target weight between four and eight weeks, and the slower growing Muscovy ducks at seven to ten weeks (Makagon & Riber, 2022). During this short period of time prior to slaughter, some may never leave the barren, crowded grower barn nor experience the joy of paddling and submerging in an adequate water source. Ducks are waterfowl who have adapted to spend a large part of their life in and around water, an environment difficult, if not impossible, to replicate in confined indoor environments.

Stocking density greatly influences space use, social behaviours and environmental conditions such as litter and faecal matter, moisture and noise. While all these issues raise obvious welfare concerns for the confined ducks, there were also economic implications for productivity which ironically often provided greater motivation to address the situation. Behavioural, clinical and production-based indicators all point towards problems with poor feather quality, foot pad inflammation, reduced feed intake, slower weight gain, increased mortality and abnormal behaviours such as pecking and excessive aggression (Yang et al., 2020). Physiological stressors increase plasma corticosterone levels and reduce immunity, all creating a financial burden for producers, and more importantly, a psychological burden on ducks unable to carry out important duck tasks like exploring, foraging, preening, bathing and socialising with chosen friends.

Ducks need to be around water to preen and clean themselves (Compassion in World Farming, 2021). Without access to adequate water, they can show compensatory behaviours of increased and excessive preening if and when they finally access water, as if trying to make up for lost time. If completely deprived, abnormal behaviours such as head shaking and stereotypic relentless preening signify severe psychological distress as well as physiological distress associated with lack of sweat glands for thermoregulation and potential heat stress. The lack of global standardised legal requirements to provide commercially farmed ducks with water for bathing or swimming means some may be provided with troughs, while others may be limited to drinking water from ball-bearing "nipples" (Jones et al., 2009). Nipple drinkers can in no way satisfy a duck's innate motivation to bathe, further compounding the stress associated with overcrowding and boredom.

These factors can also be precursors to aggression with negative implications for welfare. Rather than addressing the obvious social and environmental causes of bullying, feather pecking and cannibalism, beak trimming—often within the first three weeks of life—became a legal option in some regions. The economic losses from damage to plumage and skin, or death of birds, are a flawed rationale for the pain and suffering inflicted when removing the sensitive rim full of nerve endings at the front of the upper bill. Whether cold-cutting with scissors, hot-blade cutting with cautery or tip-searing with a cautery blade, the potential for acute and chronic pain and feeding difficulties has made bill-trimming a controversial ethical issue (El-Shafaei et al., 2017). Observations of trimmed ducks suggest that they engage in guarding behaviours such as tucking their bill under their wing, a sign of pain and suffering for the sake of productivity.

Outdoor access has been cited as a logical solution to counter aggression and the need for bill trimming (Makagon & Riber, 2022). Flock densities tend to be lower in pasture systems compared to indoor commercial systems, and access to natural light and bathing water expands behavioural opportunities such as foraging, preening and socialising. However, it does come with risks such as predation of domestic ducks who may not be as vigilant as their wild ancestors, and culling associated with avian flu, especially where the presence of wild waterfowl increases the likelihood of infection from this ongoing threat.

Whether confined in intensive settings or raised in pasture systems, slaughter is an inevitable outcome for commercial meat ducks and eventually for breeder ducks. A duck's natural life span can be up to 12 years, but no commercial duck will live to see that age when roughly caught, shoved into crates, irrespective of broken bones and wings, and sent off to slaughter. Arriving at the slaughterhouse, they are usually stunned prior to slaughter by exposure to controlled atmosphere or electrical stunning (Friedman et al., 2021). The exception is those countries that allow non-stunned religious slaughter, which requires a ventral neck incision through skin, muscle, trachea, oesophagus, sensory nerves, connective tissue, carotid arteries and jugular veins. All ducks have an extensive connective tissue sheath surrounding the jugular vein and carotid arteries, but Pekin ducks have an added complexity with a small alternative circulatory supply to the brain that can make it difficult to ascertain how long unstunned ducks take to lose consciousness. Still being conscious when immersed in the scalding hot water of the defeathering tank risks unimaginable pain and distress.

For ducks who are stunned before slaughter, there are further ethical concerns given their unique duck attributes. In some countries, gas killing devices are used, but ducks are particularly resistant to the carbon dioxide used in these devices due to their breath-holding capabilities refined by centuries of sticking their heads underwater and bottoms in the air. Carbon dioxide inhalation is unpleasant, prompting some ducks to utilise this strategy when exposed to the higher concentrations needed for extended periods of time before loss of consciousness (RSPCA, 2022).

Hanging a duck upside down as they struggle futilely over an electrical water bath is another stunning method with serious ethical and welfare implications. The shackled duck can experience painful leg dislocations, distress and suffering, especially where considerable variation in electrical parameters between individual birds affects stun efficiency (Friedman et al., 2021). In theory, a duck is stunned and killed simultaneously as their head enters the electrified water and the current flows from the submerged electrode in the water bath through their head, body and legs to the metal restraining shackles and finally to the earthed rubbing bar. However, ducks have long necks allowing them to bend their neck backwards and parallel to their back when hung upside down, thus keeping their head above the electrified water bath. Should their chest contact the electrified water first rather than the brain, loss of consciousness may not occur. Similarly, first contact by wing or beak can result in painful pre-stun shocks that prevent effective stunning. Both stunned and non-stunned methods specify that the duck must be clinically dead before entering the scalder, but during mass slaughters, this cannot be guaranteed resulting in unacceptable welfare issues. The cruelty of treating sentient animals in this way is increased exponentially when considering the inhumane treatment of foie gras ducks in the weeks leading up to slaughter.

Foie gras production

In some countries, ducks have food crammed down their throats to produce the prized delicacy of foie gras for human consumption. Foie gras means fatty liver, a luxury food item produced by concentrated force-feeding of ducks and geese. So cruel is the process of creating bloated livers, an increasing number of countries have banned or are in the process of banning its production. By the twentieth century, ducks had become the primary source of foie gras, particularly the sterile hybrid Mulard duck, which could be easily bred year-round for a steady supply of fattened livers (Kolpas, 2021). Mulard ducks are a hybrid between Pekin and Muscovy ducks. With both parents sensitive to environmental stressors, reports of higher basal levels of stress hormone corticosterone in their offspring associated with increased panic responses and fear of humans make close confinement and force feeding particularly torturous (Arnaud et al., 2008).

The livers of male ducks are preferred on the premise that males are larger and better able to withstand the force-feeding schedule prior to death. In a reversal of the fate of male chicks, it is the female ducklings who are deemed useless surplus in the foie gras trade and likely to be thrown into grinders for animal food (PETA, 2025). The detection of avian flu in these stressed, immune weakened ducks saw French producers take the unprecedented step of including female ducks in 2022, but this was out of desperation to maintain supply.

Producing the final product for foie gras is a gruesome, painful process. Young ducks have long metal tubes, called gavages, shoved down their throat multiple times a day in order to pump their stomachs full of calorie-dense food so that their liver grows six to ten times in size in a few short weeks (Kolpas, 2021). During this time prior to slaughter, they are confined in cramped cages without access to water for bathing or immersing their heads. Panting and filthy as the digestive pressure caused by forced feeding generates extra heat, they are unable to move even when their organs start failing, and they increasingly need to be restrained to continue the force-feeding regime. Mortality rates are high among these ducks who suffer unsustainable physical and psychological harm during their short miserable lives.

Those Mulard ducks fortunate enough to be released from forced feeding will not eat for eight to 15 days, and it can take a month before normal liver function is resumed. The flawed argument that overeating to enlarge the liver is a natural adaptation in preparation for migration does not work for Mulard ducks who are not migratory, nor does it in any way justify the cruel torture of these birds to create a luxury food item. It is a sad indictment on humanity that foie gras production continues and has even been described as part of France's "official gastronomic heritage" (DeSoucey, 2016, p. 5). Positioning foie gras as a cultural symbol of a country's culinary pride and excellence turned foie gras production into a tradition rather than an act of blatant animal cruelty. As more and more countries ban its production, continued demand leaves ducks in other countries vulnerable to human greed.

FEELINGS AND EMOTIONS

Bird brains may be wired slightly different to mammal brains, but they can still feel what is happening to them. If that was not already obvious from their vocalisations and body language, the 2012 *Cambridge Declaration on Consciousness* made it very clear with the decision that humans were not unique in possessing the neurological substrates that generate consciousness—birds had them too.

Put simply, ducks can experience what is happening to them, and they are capable of being harmed by negative experiences. Stressors set in motion a complex physiological response that results in the release of glucocorticoids (Tetel et al., 2022). A primary goal is to increase blood glucose levels to prepare the duck for fight or flight, depending on the stressor. In migratory birds like Mallard ducks, glucocorticoids are also associated with nocturnal restlessness, increased feeding behaviours and fat accumulation, a process that is cruelly distorted when force feeding ducks to produce foie gras.

The perception that bird brains are somehow lesser than mammals is still hard to shake. There are some notable differences like a lack of separation of grey and white matter in birds, but they also have a higher neuron density packed into a smaller brain, so that implies shorter distances to send messages between connected neurons (Rose, 2022). Their well-developed midbrain includes a large visual centre that develops early during the embryonic stage and supports vision as a primary sense (Yang et al., 2022).

They quickly become bored and frustrated when kept in environments with no enrichment, which can lead to aggression towards one another. To explore how ducks develop and think, researchers Antone Martinho and Alex Kacelnik (2016) exposed newborn ducklings to a variety of objects, showing them pairs that were either the same or different, in characteristics like shape or colour. When later shown completely different objects, around 70 per cent of ducklings approached the pair that had the same relation they had originally seen and imprinted on—whether it was one of colour or shape, sameness or difference. The researchers concluded that from point of hatching, the ducklings' brains were prepared to pick up abstract relational properties between their sensory input and the objects' characteristics, a crucial survival skill for precocious birds needing to identify mama duck and siblings in spite of considerable differences in how they are perceived. The imprinted target object was not bound by a single criterion such as colour, shape or symmetry—or mother duck only standing still. If they were relying on a fixed image, that would mean every time mother duck submerged her head in water, or flew, or even moved behind a bush and her shape and appearance changed, the young ducklings would risk losing their mother. While these findings have been open to some differences in interpretation, they do suggest that ducklings have some capacity to form complex multidimensional representations of their target object, either at time of learning or subsequently through generalisation.

Relational matching behaviour has been demonstrated in some primates, such as monkeys, apes and human infants, and a few other birds, like parrots and crows, none of whom are confined and slaughtered like ducks in intensive farm settings. These findings are a timely reminder that abstract relational thinking may no longer be the sole domain of humans and certain privileged animals and, in fact, may reside in the animals that humans choose to commodify.

Even before ducklings hatch, their parents are communicating with them, and siblings are talking to each other from their eggs (Ducks Unlimited Canada, 2025). Ducks also do a lot more than just quack—there are whistles, coos, purrs, grunts, barks and honks at various sound levels as they interact socially and express their emotions. They call for their babies or their mate, warn them of predators or threats or generally chatter with appropriate body language. Researchers Jenna Schober and colleagues (2024) identified up to 16 different vocalisations in their study among Pekin ducks, including unpublished observations of four additional egg laying vocalisations, and noted that males and females differed in their repertoire of calls.

With chirps and quacks, bobbing heads and wagging tails, it is hard to resist smiling at the sight of a secure and happy duck showing affection to family members and friends or playfully chasing each other round, stopping to bill-wrestle or preen each other's feathers. Similarly, seeing a mother duck lead her brood across a dangerous road inevitably triggers a nurturing response in some people who rush to stop traffic so the family may pass. This is not surprising, given zoologist, ethologist and ornithologist Konrad Lorenz's (1943) observation that infantile physical features such as large foreheads, big eyes, bulging cheeks and soft contours prompted greater attachment behaviours in humans. Fluffy cute ducklings can prove irresistible while young, but they do grow up. If given the opportunity, ducklings stay with mothers for 40–60 days before flying off to make their own way in life. Their mothers protect them from rain, hail, snow, heat, wind and predators as they teach them what to eat and how to survive.

These affectionate sentient birds can also feel pain, isolation and grief, a growing reality for many ducks confined in industrial farming systems, abandoned like Neil Diamond and Garry or raised alone in backyard settings. Ducks reared in isolation risk negative impacts on their emotional wellbeing. This can happen when ducklings are sold at live bird markets, through online marketplaces or raised in school hatchling programmes and adopted by well-meaning people who may not understand their unique needs. Ducks need access to sufficient clean water to fully submerge and perform all their fun and necessary water-related activities such as preening, bathing, floating and just hanging out with friends and partners.

Online classified advertisements selling ducklings for a few dollars each for use as “pets or meat”, accompanied by pictures of fluffy ducklings, remain a threat to the adult duck’s wellbeing when their specific needs cannot be met. Some of these ducks are dumped when they reach adulthood and are of no further use or have become problematic, leaving them vulnerable to predator attack, injury or death.

Sheri, Jody and Bumblebee, unique personalities—Woodstock Farm Sanctuary, USA

Kiwi, Sheri, Jody and Bumblebee arrived at Woodstock Farm Sanctuary in High Falls, New York, USA, for a variety of reasons and in different states of health and wellbeing. Some were dumped, others were part of school hatchling programmes, while still others lived with physical limitations from selective breeding. Once safe from the physical and emotional stressors of their earlier lives, each revealed their unique personalities and preferences in the bonds they formed and the resilience they showed when adapting to any chronic health conditions. The duck pond at Woodstock facilitated friendships and lifelong partners, some of whom would remain together for over a decade.

Woodstock Farm Sanctuary began in 2004 outside of Woodstock, New York (Woodstock Farm Sanctuary, 2025). Like many farmed animal sanctuaries, growing demand saw them relocate to 150 acres in the hills of High Falls, New York, in 2015. With a vision and mission to rescue farmed animals, give them lifelong care and sanctuary and educate about the harms of animal farming, Woodstock has provided thousands of animals the chance to live their lives with dignity and respect. The ducks, chickens, geese, cows, pigs, sheep, goats and turkeys who currently call Woodstock Farm Sanctuary home are testament to the sanctuary’s ongoing advocacy for animal rights in alliance with other social justice movements.

Sheri

When Woodstock resident Sheri the Muscovy duck fractured her femur, there was no hesitation in seeking immediate—and expensive—surgery. The economic reality of intensive duck farming settings is that a single duck's life does not warrant expensive medical intervention, making an injured duck like Sheri a disposable commodity. However, to the team at Woodstock Farm Sanctuary, Sheri was someone, not something, and deserving of all their care and respect at this crucial time.

Sheri had been found abandoned in a park with only her trimmed beak giving any indication of her background. When large numbers of ducks are confined together, increased stress levels can result in feather pecking and potential injuries or death to themselves or other birds. Instead of addressing the cause of this problem, beak trimming is sometimes used to remove the tip of a bird's beak and blunt it sufficiently to reduce the negative impacts of severe feather pecking. This is despite evidence that beak trimming has the potential to cause acute and chronic pain (El-Shafaei et al., 2017) and that it may in fact increase feather pecking when the bird experiences frustration at being unable to grasp feathers and continues pulling to complete the action (Marchewka et al., 2019). Sheri also had an infected foot, which can arise from factors such as inappropriate flooring and sanitation or lack of swimming water, both of which suggest that Sheri may have come from less-than-ideal circumstances.

Like many animals who reach the safety of a farmed animal sanctuary, Sheri soon left her past traumas behind as she enjoyed time with her duck friends swimming in their pond. After the surgery on her femur, Sheri was confined inside while she healed. When the big day arrived for Sheri to again venture outside, she was meant to start in a yard on her own for safety. Sheri however had other ideas. After watching her attempts to scale the shade sail separating her from the friends she could hear quacking excitedly, the team relented. The risk of Sheri hurting herself as she tried to break out far outweighed any risks in the main duck barn, and so Sheri got what she wanted and rejoined her group a little earlier than planned. Watching her friends rush to her side, bobbing heads and wagging tails as they chattered non-stop, it seemed that the group was keen to catch up on all the news of the past few weeks. Some might call this interpretation an anthropomorphic projection, but it did provide a valuable means to make sense of what Sheri needed at that point in time. Ethologist Carl Safina commented that if we can recognise that an animal is hungry or thirsty, why cannot we recognise that they are happy or joyful when acting that way? (Safina, 2015). Adopting this perspective makes anthropomorphism a powerful linguistic mechanism to interpret the thoughts and feelings of animals such as Sheri and the desperation she demonstrated to reunite with her friends.

Jody

Jody the Pekin duck was also found abandoned, alone on a lake not too far from the sanctuary. Reports initially cited two ducks, but only Jody could be found with a large wound on her neck suggestive of a fox or dog attack and offering some clue as to the fate of the second Pekin duck. It also meant that Jody likely witnessed the death of her friend, a devastating sight for a duck capable of bonding for life. Jody was found limp in a clump of grass, but the trust she showed her rescuers as she cuddled in their arms was testament to the intuitive recognition of safety that many animals display.

Jody's journey to health was only just beginning. The horrific external wound masked even greater internal damage, necessitating a diet of nourishing smoothies filled with blueberries and pellets as her body healed. Jody would not have survived alone much longer, nor had the opportunity to become a loyal friend to many other ducks at Woodstock. With the same resilience shown by Sheri, Jody recovered from the earlier traumas of abandonment, death of her friend and a life-threatening wound once in the safety and security of Woodstock Farm Sanctuary.

Bumblebee and the "bugs"

Sheri and Jody arrived as adults whose past lives remained a mystery, but this was not the case for Bumblebee, a young Pekin duck. Bumblebee and her siblings Cricket, Beetle, Ladybug, Firefly and Caterpillar, collectively known as the "bugs", were part of a school hatchling project. School hatchling programmes present an ongoing ethical challenge as they promote speciesism under the guise of education. Young children are taught that non-human animals are commodities to be hatched in the classroom as a project and then discarded when no longer useful.

After being rescued by a Humane society who succeeded in dissuading the school from continuing this decades-long practice, all six hatchlings were relocated to Woodstock. As the youngest baby birds ever accepted into the sanctuary, these little fluff balls with their irresistible paedomorphic features captivated the hearts of all those who would spend countless hours nurturing them to adulthood. The inseparable ducklings learned the joy of natural behaviours as they moved from the medical barn to their own house and had the freedom to swim together in the sun and cuddle up during their nap time, activities unlikely to have been fulfilled in their previous life trajectory.

However, Bumblebee could not escape the legacy of human selective breeding. Much like chickens and turkeys, Pekin ducks have been bred to grow rapidly in preparation for early slaughter. As their skeletons buckle under this unnatural growth, they face chronic pain from joint dislocations and foot and leg fractures during their shortened lifecycle. Bumblebee escaped early slaughter, but she was left with the legacy of these abnormalities and resultant disabilities. Bumblebee and her Pekin duck friend Toots shared numerous vet visits as they battled respiratory problems and foot issues during the ongoing treatment regime denied to Pekin ducks in the intensive farm system. For both, medical care would be lifelong and expensive, something that all farmed animal sanctuaries accept as they offer sanctuary to those animals damaged by human use and abuse.

ETHICAL IMPLICATIONS

Animal agriculture economics positions animals as things that are intermediate production goods rather than individual sentient beings (Espinosa & Treich, 2024). From an economist's perspective, the intrinsic value of animal welfare is

Figure 11.2 Bumblebee and the "bugs" arrived as tiny fluff balls with irresistible paedomorphic features. (Image supplied by Woodstock Farm Sanctuary and reproduced with their permission.)

inextricably linked to the extrinsic value of how it might benefit the quantity or quality of output. The tension between animal lives and economic implications is evident in this headline immediately following an outbreak of avian influenza in Australia in 2024:

> Australia is recovering from its worst outbreak of bird flu, which has led to the deaths of 2 million chickens across 16 farms — leading to egg shortages and economic pain for farmers.
>
> *(Tippet, 2024)*

Overshadowing the horror of the death of two million sentient beings is the greater impact of egg shortages for consumers and economic losses for farmers.

Ongoing news items reiterated the increased cost of eggs and adopted the less emotive language of quarantine and culling as opposed to death and killing, followed by "re-stocking" once the immediate danger was past. Although this headline refers to chickens not ducks, it reflects a common anthropocentric perspective adopted for farmed poultry in a range of situations. The language moves the reader beyond the horror of death on an unimaginable level to focus on the direct impact to themselves. In this scenario, the welfare of a single bird was lost within the economic impact and the psychic numbing that can provide a flawed protective mechanism when confronted with death on such a massive scale. This strategy could equally apply to the implicit knowledge that ducks farmed for their flesh or foie gras can be kept in horrific conditions, highlighting the paradox that allows knowledge of the mass confinement of ducks to co-exist with a person's declared concern for animal welfare.

A research study of consumer welfare attitudes to laying ducks in Taiwan identified two stages of decision-making and thinking processes in relation to animal welfare: consumers considered whether they had to pay extra for animal welfare products, followed by how much extra they would have to pay. Believing that their actions would make a difference was also a factor when considering ethical consumption of animal products (Yang & Wang, 2025). As evidence accumulates that raising awareness of animal welfare can make a difference when providing a rationale to combat cognitive dissonance, farmed animal sanctuaries provide an integral component in both animal welfare and animal advocacy as they show that every animal's life matters.

Like other farmed animal sanctuaries, Woodstock Farm Sanctuary combats the commodification and objectification of farmed animals by focusing on their individuality. Each animal, irrespective of their perceived economic value in the broader farming community, has the right to care, respect and any medical interventions necessary to achieve quality of life. Contrary to rhetoric around the misuse of expensive resources to mend animals such as ducks with low extrinsic value, all animals entering a farmed animal sanctuary are granted "personhood" and equity in care. When a duck named Kiwi arrived at Woodstock Farm Sanctuary, it was clear that the journey to allow her a chance at life would be long, hard and expensive. But it was also a journey that inspired a global following, as the brave little duck proved that every life is important.

Kiwi, a life worth saving—Woodstock Farm Sanctuary, USA

Kiwi the Pekin duck arrived at Woodstock Sanctuary in 2020. Originally part of a school hatchling project, Kiwi was meant to return to the supplier when she became too big for the classroom. This did not eventuate, and Kiwi was adopted by a kind family until a predator attack left her with spinal damage and unable to walk or stand. Like Jody, Kiwi most likely witnessed the traumatic death of her friend during the attack.

Kiwi met all the criteria for the human concept of "disabled" and at risk of the "better off dead" narrative that is ubiquitous among disabled farmed animals (Taylor, 2020). Fortunately, Kiwi found her way to the sanctuary where this narrative quickly changed to one of resilience and hope when the brave little duck made it very clear that she wanted to move and enjoy this second chance at life. Two custom-made sets of wheels—one for indoors and one for outdoors—plus physical therapy on her legs, swimming in a tub for an hour a day, and Kiwi's unquenchable fighting spirit turned the little duck into a global superstar.

The unforgettable video of Kiwi zipping down the muddy road on four wheels behind her goat friend Remy on his wheels, cheered on by a squad of goats, showed the little duck's determination to not only keep up, but to live life to the fullest. Remy went on to receive prosthetic legs, another example of how a disabled animal born into an uncertain future on a meat and dairy farm could become somebody with a future. Whether Kiwi and Remy recognised that they moved differently than other members of their species remains unknown, but both demonstrated survival, adaptation and apparent enjoyment in the freedom that human-made wheels afforded them that day.

Watching Kiwi's unique swimming style as she navigated the wading pool with one leg waving in the air also raised questions as to whether the other ducks recognised their friend was different. The human concept of ableism seemed to play no part in Kiwi's new world of friendships and care. Soon after her arrival, Kiwi struck up a friendship with Donut, an older duck with a traumatic past and leg issues, and the two of them could often be found lying on the grass in the sunshine. Best friend Gimli the Pekin duck who also experienced mobility issues after growing too big too quickly, loved splashing around with Kiwi during water therapy, each seemingly oblivious to the other's disability as they enjoyed their time together. Breakfast and lunch dates were the norm for these two friends, although Kiwi's medical check revealed all the treats associated with the extra love and attention needed cutting back and replacing with a diet of vegetables—something that prompted strong vocal complaints from the feisty little duck. As ducks are such social animals, friends play a crucial role in any rehabilitation programme, and Kiwi was never short of duck and human friends alike.

Kiwi was so much more than a school project or a disabled animal who would have been better off dead, nor was her worth ever measured in terms of productivity and ability. The compassion and respect she received at Woodstock Farm Sanctuary ensured that she enjoyed the happiness and quality of life so deserving of every sentient animal.

In May 2025 a Facebook post advising that Kiwi had passed away prompted an outpouring of grief from all those who had followed Kiwi's five years of incredible life at Woodstock Farm Sanctuary.

Figure 11.3 Kiwi's wheels allowed her to move around the sanctuary. (Image supplied by Woodstock Farm Sanctuary and reproduced with their permission.)

REFERENCES

Animals Australia (2025). *Save native ducks from 'recreational' slaughter.* https://animalsaustralia.org/our-work/shooting-and-hunting/duck-shooting-victoria/

Arnaud, I., Mignon-Grasteau, S., Larzul, C., Guy, G., Faure, J.M. & Guémené, D. (2008). Behavioural and physiological fear responses in ducks: Genetic cross effects. *Animal, 2,* 1518–1525. https://doi.org/10.1017/S1751731108002784. PMID: 22443910.

Cambridge Declaration on Consciousness. (2012, July 7). Written by Philip Low and edited by Jaak Panksepp, Diana Reiss, David Edelman, Bruno Van Swinderen, Philip Low and Christof Koch. University of Cambridge.

Challenger, M. (2021). *How to be animal. A new history of what it means to be human*. Canongate Books Ltd.
Compassion in World Farming (2021). *Scientific briefing on caged farming. Overview of scientific research on caged farming of laying hens, sows, rabbits, ducks, geese, calves and quail*. https://www.ciwf.org.uk/media/7446651/150603_ciwf-february-2021-scientific-briefing-on-caged-farming.pdf
DeSoucey, M. (2016). What can we learn from liver? *Princeton Scholarship Online*, 1–26. https://doi.org/10.23943/princeton/9780691154930.003.0001
Ducks Unlimited Canada (2025). *A hen's journey*. https://www.ducks.ca/a-hens-journey/
EFSA Panel on Animal Health and Animal Welfare (AHAW Panel), Nielsen, S.S., Alvarez, J., Bicout, D.J., Calistri, P., Canali, E., Drewe, J.A., Garin-Bastuji, B., Gonzales Rojas, J.L., Schmidt, C.G., Herskin, M., Michel, V., Miranda Chueca, M.Á., Padalino, B., Roberts, H.C., Spoolder, H., Stahl, K., Viltrop, A., Winckler, C., Berg, C., … Velarde, A. (2023). Welfare of ducks, geese and quail on farm. *EFSA journal. European Food Safety Authority, 21*(5), e07992. https://doi.org/10.2903/j.efsa.2023.7992
El-Shafaei, H.E., Sharaf, M.M., Rashed, R.R., & Elkazaz, S. (2017). Consequences of bill trimming on behavior, welfare and performance of Muscovy ducks. *Alexandria Journal of Veterinary Sciences, 55*, 124–128. https://doi.org/10.5455/ajvs.282690
Espinosa, R. & Treich, N. (2024). Beyond anthropocentrism in agricultural and resource economics. *Australian Journal of Agricultural & Resource Economics, 68*(3), 541–566. https://doi.org/10.1111/1467-8489.12560
Friedman, A., Dalla Costa, F.A., Dalla Costa, O.A., Godsell-Ryan, A., & Gibson, T.J. (2021). Time to loss of behavioral and brainstem responses of ducks following non-stunned slaughter. *Animals, 11*, 3531. https://doi.org/10.3390/ani11123531
Herrera-Castillo, C.M., Geiger, M., Núñez-León, D., Nagashima, H., Gebhardt-Henrich, S., Toscano, M., & SanchezVillagra, M.R. (2022). Skeletal variation in bird domestication: Limb proportions and sternum in chicken, with comparisons to mallard ducks and Muscovy ducks. *PeerJ, 10*, e13229. https://doi.org/10.7717/peerj.13229
Jiang, F., Lin, R., Xiao, C., Xie, T., Jiang, Y., Chen, J., Ni, P., Sung, W.K., Han, J., Du, X., & Li, S. (2021). Analysis of whole-genome re-sequencing data of ducks reveals a diverse demographic history and extensive gene flow between Southeast/South Asian and Chinese populations. *Genetics Selection Evolution, 53*(1), 35. https://doi.org/10.1186/s12711-021-00627-0. PMID: 33849442; PMCID: PMC8042899.
Jones, T.A., Waitt, C.D., & Dawkins, M.S. (2009). Water off a duck's back: Showers and troughs match ponds for improving duck welfare. *Applied Animal Behaviour Science, 116*(1), 52–57. https://doi.org/10.1016/j.applanim.2008.07.008
Kolpas, N. (2021). *Foie Gras. A global history*. Reaktion Books.

Lorenz, K. (1943). Die angeborenen Formen möglicher Erfahrung. *Zeitschrift Für Tierpsychologie*, 5, 235–409.

Makagon, M. M., & Riber, A. B. (2022). Setting research driven duck-welfare standards: A systematic review of Pekin duck welfare research. *Poultry Science*, 101(3), 101614. https://doi.org/10.1016/j.psj.2021.101614

Marchewka, J., Vasdal, G., & Moe R.O. (2019). Identifying welfare issues in turkey hen and tom flocks applying the transect walk method. *Poultry Science*, 98(9), 3391–3399. ISSN 0032-5791. https://doi.org/10.3382/ps/pez211

Martinho III, A. & Kacelnik, A. (2016). Ducklings imprint on the relational concept of "same or different". *Science*, 353(6296), 286–288. https://doi.org/10.1126/science.aaf4247. PMID: 27418508.

PETA (2025). *Foie Gras: Cruelty to ducks and geese*. https://www.peta.org/issues/animals-used-for-food/factory-farming/ducks-geese/foie-gras/

PETA Australia (2024). *Government rejects its own inquiry – Duck hunting in Victoria to continue*. https://www.peta.org.au/news/government-rejects-its-own-inquiry-duck-hunting-in-victoria-to-continue/

Rose, J. (2022). The avian brain. *Current Biology*, 32(20), R1076–R1079. https://doi.org/10.1016/j.cub.2022.07.072

RSPCA (2022, March 22). *Duck farming: Animal welfare considerations*. https://kb.rspca.org.au/wp-content/uploads/2018/11/Duck-farming-Animal-welfare-considerations.pdf

RSPCA (2024). *Farmed duck welfare needs our attention*. https://www.rspca.org.au/latest-news/blog/farmed-duck-welfare-needs-our-attention/

Safina, C. (2015). *Beyond words: What animals think and feel*. Henry Holt and Company, LLC.

Schober, J.M., Merritt, J., Ulrey, M., Yap, T.Y., Lucas, J.R., & Fraley, G.S. (2024). Vocalizations of the Pekin duck (*Anas platyrhynchos domesticus*): How stimuli, sex, and social groups affect their vocal repertoire. *Poultry Science*, 103(7), 103738. https://doi.org/10.1016/j.psj.2024.103738

Sun, H.X., Guo, R.B., Gu, T.T., Zong, Y.B. Xu, W.W., Chen, L., Tian, Y., Li, G.Q., Lu, L.Z., & Zeng, T. (2024). Investigating the correlation between phenotypes, adrenal transcriptome, and serum metabolism in laying ducks exhibiting varying behaviours under the same stressor. *Animal*, 18(11), 101343. https://doi.org/10.1016/j.animal.2024.101343

Taylor, S. (2020). Animal crips. In S. Jenkins, K. Struthers Montford, & C. Taylor, C. (Eds.), Disability and Animality (1st ed.) (pp. 13–34). Routledge. https://doi.org/10.4324/9781003014270-3

Tetel, V., Van Wyk, B., & Fraley, G.S. (2022). Sex differences in glucocorticoid responses to shipping stress in Pekin ducks. *Poultry Science*, 101(1), 101534. https://doi.org/10.1016/j.psj.2021.101534

Tippet, H. (2024, October 1). Australia's bird flu outbreak has claimed the lives of 2 million chickens, but experts fear the worst is yet to come. *ABC News*. https://www.abc.net.au/news/2024-10-01/bird-flu-h5n1-australia-chicken-native-birds-egg-farmers-csiro/104413006

Woodstock Farm Sanctuary (2025). *Our history*. https://woodstocksanctuary.org/history

Worldostats (2025). *Duck population by country 2025.* https://worldostats.com/duck-population-by-country-2025/#:~:text=According%20to%20the%20Food%20and%20Agriculture%20Organization%20%28FAO%29%2C,commercial%20purposes%2C%20and%20stock%20ducks%2C%20maintained%20for%20breeding

Yang, C., Yu, C., Zhang, Z., Qiu, M., Du, H., Li, Q. & Xiong, X. (2020). Research progress on effects of stocking density on performance, health and welfare of duck. *Agricultural Biotechnology*, 9(4), 111–114, 119.

Yang, S., Wu, J., Shi, Y., Huang, Y., Zhang, Y., & Chen, Q. (2022). Development of the blood–brain barrier in ducks. *Microscopy and Microanalysis*, 28(2), 504–514. https://doi.org/10.1017/S143192762200006X

Yang, Y. & Wang, C. (2025). The determinants of willingness to pay for improving the animal welfare of laying ducks: The case of Taiwan. *International Food and Agribusiness Management Review* (published online ahead of print 2025). https://doi.org/10.22434/ifamr1018

12

Alpacas and llamas: Charismatic and sensitive personalities

INTRODUCTION

In October 2024, the British Monarch, King Charles III, met 9-year-old Hephner the alpaca during a walk-about in Canberra, Australia (Hill, 2024). Things did not go as planned, and Hephner found himself making world headlines when he sneezed on the king. King Charles was drawn to the white alpaca who stood out from the crowd in his suit, bow tie and a gold crown atop his fluffy head. Just as King Charles reached out to pat the regal animal, Hephner sneezed. This greatly amused the king, who laughed off the unexpected turn of events.

With their soft fluffy fleeces and eyelashes sweeping over soulful eyes that belie their cheekiness, alpacas and llamas fulfil many roles after spreading around the world from their origins in South America. Hephner is a support animal for charities and nursing homes, but the gentleness that makes Hephner a support animal has also contributed to llamas and alpacas being cruelly exploited for food, clothes and entertainment.

While both alpacas and llamas originated in South America, growing numbers in North America, Australia, New Zealand, Canada, United Kingdom, Ireland and mainland Europe has highlighted the need for greater research into husbandry practices and their emotional wellbeing (la Lama & Villarroel, 2023). In 2010, there was an estimated 36,000 alpacas and llamas in the United Kingdom (Duncanson, 2023). By 2023, that number had risen to more than 58,000 alpacas and llamas. Australia is home to 175,000 alpacas and llamas, all originating from a small herd of 250 brought into Australia in 1858, while New Zealand has 15,000 alpacas and llamas. The USA has over 400,000 alpacas and llamas, dating back to a small import in 1984, but all these are only a fraction of the Andean countries of South America, where an estimated 3.8 million llamas and four million alpacas reside (la Lama & Villarroel, 2023). Peru has the largest number of the world's

DOI: 10.1201/9781003518853-13

alpacas, while Bolivia has the most llamas. The majority live with small producers in mixed herds of llamas, alpacas, sheep and cattle where their meat, fibre, manure and transport play a crucial role in the livelihood of Andean farming communities. As the versatile alpacas and llamas gained global popularity, their commodification and inclusion in intensive farming systems with accompanying restriction of natural behaviours has raised increasing concern for their welfare.

SOME BACKGROUND HISTORY

Alpacas and llamas are Camelids and members of the biological family *Camelidae*, which also includes their wild relatives, the guanaco and vicuña, as well as camels. Evidence of the family *Camelidae* dates back around 40 million years in North America. Some family members crossed the land bridge in place at that time to Eurasia and Africa and evolved into the dromedary and Bactrian camels. Others migrated south to become domesticated llamas and alpacas, and wild or semi-wild guanacos and vicuñas (Duncanson, 2023). Any family members left in North America eventually became extinct.

Within the family, *Camelidae* are two genera: the *Vicugna* genus which includes the wild species vicuña and the domestic alpaca, and the *Lama* genus which includes the wild species guanaco and the domestic llama. Mitochondrial analyses have confirmed that llamas were domesticated from guanaco subspecies and alpacas from vicuñas, although there was also interbreeding among llama and alpaca (la Lama & Villarroel, 2023). The llama is larger than the alpaca, but an easy differentiation is in their ears. Llamas have long, banana-shaped ears which they use to express their feelings. The alpaca's triangular ears are similarly expressive, but considerably smaller in size (PETA, 2025).

The guanaco and vicuña evolved from natural selection in the wild and remained untamed among the indigenous peoples and "protected" by the mountain deities, but the alpacas and llamas did not. They followed a different trajectory, becoming a co-creation of nature and the requirements of Andean inhabitants who claimed ownership of these animals (Vilá & Arzamendia, 2022). Transforming the guanacos and vicuñas into llamas and alpacas changed the lives of humans living in the Andes and prompted the transition from hunting to herding and village life. With nail-covered plantar pads on their feet instead of hooves, alpacas and llamas could safely traverse the steep slopes of the dry high-altitude Andean steppes of South America, making them valuable possessions for any humans living there.

For a time in history, their social value as respected material and symbolic resources deteriorated to being considered no more than those "animals from the Andean Indians" and of little consequence (Vilá & Arzamendia, 2022, p. 709). This coincided with the arrival of the Spanish Conquistadors and their attempts to introduce cattle, sheep and goats into the region. However, they regained their position as important contributors to Andean indigenous and local communities when their ecological significance as low impact grazers on arid steppes became increasingly evident. Alpacas were of particular importance to villagers occupying high-altitude wetlands and grasslands through the provision of high-quality fibre and meat, while llama caravans were common as a means of transporting goods across the Andes.

ALPACAS AND LLAMAS IN CONTEMPORARY SOCIETY

Alpacas and llamas have followed many different pathways from their early life in the Andes but with minimal changes in physiology compared to other domesticated animals. Described as atypical examples of domestication, their bodies, including skull shape, did not succumb to major modifications arising from human selective breeding. They remained intensely curious, with a strong herd instinct in which a single dominant male patrolled and protected the group until deposed by another male (Duncanson, 2023). This protective trait saw some llamas and alpacas used with sheep flocks to protect young lambs. An Australian study reported alpacas chasing predators away, rounding sheep into small areas and placing themselves between the sheep and threats (Mathews et al., 2024). As herd animals, alpacas and llamas do not fare well on their own, and while they may bond with other species, two or more living together is preferable (Australian Alpaca Association, 2020). The implications of social isolation, especially as these gentle animals became popular exotic companion animals in Western nations, was evident in increased heart rate, depression and lethargy.

There is no doubt that alpacas and llamas are cute. As with many young animals, the huge eyes and soft features of infant crias could elicit a nurturing instinct in humans. Ideally, crias remain with their mother for up to six months before being ready for weaning. This allows the mother time to teach her infant all the skills needed to be an alpaca or llama, including grazing pasture identification and predator avoidance techniques. However, crias are curious and make contact with humans easily. Raising a cute cria as a pet and excessively fussing over them from an early age may prevent them learning the right essential skills and risk the development of serious abnormal behaviours and problems in later life. The potential for "mal-printing", when a growing cria cannot tell the difference between a human and their own species, can be appealing in a youngster when they are easy to handle, but not so cute when the infant turns into an adult. The mal-printed adult who believes the human is their own species can display excessively familiar behaviour and even domination. Originally given the dramatic label "Berserk Male Syndrome", this was amended on the realisation that not just males were vulnerable to the negative implications of over-petting and handling (Duncanson, 2023). Now labelled "Aberrant Behaviour Syndrome", outcomes of this excessive human intervention could be aggression to humans and animals alike, including rearing up, chest butting, running towards a person with head lowered and biting. Once this psychological and behavioural condition became entrenched, rehabilitation or retraining was unlikely.

The normally gentle nature of llamas and alpacas also saw them utilised in a range of human interaction activities, including petting zoos, alpaca and llama trekking, and as therapy animals among stressed students before exams, retirement home residents and rehabilitation centre patients (PETA, 2025). The emotional support provided during alpaca and llama therapy is widely promoted, but as with all forms of animal-assisted therapy, there is a responsibility to ensure that the animal's needs are met and their experiences are positive. Where an animal is placed in a position to support human mental health, the reciprocal negative impact on the animal's emotional wellbeing and the risk of inter-species emotional contagion through ongoing exposure to negative human emotions becomes a risk. Placing complex animals with emotional capacity into settings specifically focused on human physical or psychological wellbeing has ethical implications, especially within the context of the mental experiences of Domain 5 of Mellor's Five Domains (Fine & Griffin, 2022). This domain states that animals must be provided with opportunities to achieve a positive mental state, which can be difficult if lacking agency in human interactions. There is an expectation that therapy animals remain compliant and docile in these settings, something that can change across time if owners lack sufficient knowledge about their physical and psychological requirements, or the animals become overly distressed. This can result in situations of abuse, neglect or abandonment, with the lucky few relocating to farmed animal sanctuaries.

Fibre production

On a larger scale, fibre production from the fleece of alpacas and llamas has been linked to a range of welfare issues in both Western countries and South America. In Europe, small farm holders breed alpacas who have dense fleece with very fine fibre, ideally all over the body, neck and head (Gunsser, 2013). Dense fleece around the head can lead to reduction in their field of vision and even "wool-blindness", a major stressor for this gregarious flight animal if unable to observe their surroundings. Where inexperienced owners fail to identify signs of stress associated with their animals' eyes partially or fully covered, this can have a flow-on effect to their overall welfare including reproductive capacity and the pre-existing potential for high embryonic loss during pregnancy (Montelli et al., 2019). Congenital deafness is evident among white llamas and alpacas with solid blue eye colour rather than pigmented eyes, a deficit that is selectively bred to meet global demand for white fleeces. These fleeces can be dyed to any desired colour, making the pure white alpaca or llama, irrespective of potential deafness, the most economically valuable animal for wool production (Jost et al., 2020). Understanding these charismatic but sensitive animals becomes an imperative as their global popularity increases.

The vast majority of llamas and alpacas reside in South America and occupy an important position in the livelihood of small producers in Andean farming communities. However, up to 20 per cent of these animals are confined to medium-sized farms and business enterprises, leaving them at risk of experiencing the same welfare issues and deprivation of natural environments identified among more traditional farmed animals in intensive farming systems. Confinement in overcrowded conditions, lacking space to distance themselves from the communal defecation pile preferred by alpacas and llamas, and subjected to potential stressors during handling and shearing, can have serious implications for their physical and emotional welfare (Duncanson, 2023).

In 2020, People for the Ethical Treatment of Animals (PETA) published video footage of alpacas being tied down and shorn at a large privately owned alpaca farm in South America (PETA, 2024). These images showed unimaginable terror and suffering during the shearing process to obtain alpaca wool, which is considered as one of the highest quality wool fibres for yarn and knitted items. Forcibly restraining a prey animal such as alpacas on their back or side was only the start of their physical and psychological trauma. As the animals screamed, spat and vomited in fear, workers were seen grabbing their ears and working quickly and carelessly, which inevitably left some animals bleeding from deep, painful wounds. Some wounds were stitched up without pain control, leading to subsequent infection. When finished, the animals were thrown onto a concrete floor where many stood frozen in fear and exhaustion. Hearing, seeing and smelling their conspecifics' terror, alpacas awaiting their turn could be seen huddled together. With the pre-existing potential for high embryonic loss during pregnancy especially when stressed, good husbandry practice recommends not shearing pregnant alpacas or llamas during the first or last 60 days of pregnancy (Duncanson, 2023). However, the video footage included pregnant females tied down and powerless. Four years after the release of this video footage, several individuals were charged with cruelty to animals, and some major global clothing suppliers undertook to phase out alpaca wool products. This exposé and eventual outcome provide a timely reminder of the welfare risk all animals can face once they become commodified on a large scale.

Research subjects

Llamas and alpacas have another surprising contribution that makes them valuable commodities for human use—their unique antibodies (Vilá & Arzamendia, 2022). Llamas and alpacas have both the four-chain antibodies (that is, two heavy chains plus two light chains) found in most mammals including humans, as well as the more unique two chain antibodies comprising only two long chains. Advances in recombinant gene technology made it possible to extract small antibody fragments called "nanobodies" or "variable heavy homodimers" from the two-chain antibodies. These have very good physiochemical properties, including the possibility of humanisation, making them uniquely suited to deliver biologically active compounds. Llamas, with their smaller and simpler nanobodies that fit into the binding sites on a viral surface, are significant contributors to human immunology research to neutralise various types of viruses, including HIV. Their unique antibodies have proven better at latching on to receptors on the human cells than current human HIV vaccines, making them valuable in the fight against HIV as well as severe COVID-19 (Understanding Animal Research, 2025). Alpacas' antibodies have also been enlisted in the study of other human conditions such as autism, Alzheimer's disease and some cancers (Hall, 2019). Alpacas and llamas used in medical research are usually housed outdoors in farm-like conditions, and blood samples collected without the need for confinement, which can be stressful for these gregarious and curious creatures.

FEELINGS AND EMOTIONS

With origins dating back to the harsh environments of the Andes and constantly changing locations and quantities of resources, it is no surprise that alpacas and llamas have developed cognitive flexibility in problem solving. During their long gestation (approximately 340 days), cortical folding allows their brain to grow in volume and expand in surface area, no different to a range of other mammals including humans (Montelli et al., 2019). The development of the central nervous system structure progresses relatively quickly in embryos, especially the early maturation of nuclei in the hind brain to enable immediate locomotor skills in the newborn. Survival of prey animals is dependent on moving quickly after birth, and watching a newborn cria stagger to their feet within 30 minutes of birth is a testament to the complexity of a brain that processes and coordinates movement in four long wobbly legs.

While brain complexity in humans and other primates is positioned as apex, camelids in general were among the first lineage of artiodactyls (that is, even-toed ungulates, such as pigs, camels and deer) to increase their neocortical complexity in the Eocene between 56 and 34 million years ago (Balcarcel et al., 2023). The camelid brain displayed considerable differentiation of the sulcal system (grooves) and expansion of the neocortex, particularly the peripheral brain, compared to other mammals. Living in open habitats necessitated well-developed visual displays and a good sensitivity to sound, even at low frequencies, but domesticated alpacas and llamas have reduced localisation acuity compared to cattle, goats and sheep (la Lama & Villarroel, 2023). This has been attributed to morphological and size alterations of the middle ear space in the skull and a general reduction in cranial capacity as an indirect result of domestication.

As prey animals, alpacas and llamas need to hide their pain and distress from anyone or anything that could be a potential predator (la Lama & Villarroel, 2023). However, despite this façade, they are sentient animals with the capacity to suffer and feel pain via nociceptors that receive and respond to damaging stimuli with signals to the spinal cord and brain. Their brain creates the sensation of pain in the relevant body part, but to show this could leave them vulnerable to predation and so there are only subtle signs that betray their distress. This can include tachycardia (increased heart rate), bruxism (teeth grinding) and tachypnoea (rapid and shallow breathing). Where pain and stress become chronic, there can be loss of appetite, lethargy, excessive resting, avoidance of contact and abnormal posture. Severe and unbearable pain and trauma—such as that experienced by alpacas being roughly handled during shearing—cannot be disguised, resulting in violent rolling and twisting to escape. Once a certain threshold is reached, the fearful llama or alpaca may enter a state of capture myopathy, characterised by decreased blood flow of oxygen and nutrients to the tissues, thereby increasing lactic acid production and risking necrosis of skeletal muscle tissue and potential fatal renal failure.

With one simple question, eighteenth century philosopher Jeremy Bentham shifted the focus from an animal's capacity to reason to the animal's capacity to suffer, thus providing a rationale for animal welfare that remains applicable in the twenty-first century. Bentham (1789) wrote "The question is not, Can they *reason*? Nor Can they *talk*? But, Can they *suffer*?", and there is no doubt that alpacas and llamas, despite their stoicism, can suffer. They have the neuroanatomical structures and neurochemical pathways fundamental to core emotions, including fear and anger, but also happiness, love, play, sadness and grief.

Chester the alpaca had spent his entire life with his mother, Ambrose. When Ambrose died a year after the pair arrived at Furever Farm in Australia, Chester's grief was raw and heartbreaking. Biologist, ecologist and ethologist Marc Bekoff affirmed that "there is no question that animals grieve [...] like humans, animals can suffer monumentally over a separation or loss" (Bekoff, 2007, p. 63). Chester's story is one of grief, but importantly, of resilience when allowed to process this grief in his own time and manner as he reshaped his world without Ambrose.

Chester's grief and resilience—Furever Farm, Australia

When a parent dies, it can be hard to believe that someone who had always been there is gone. The human grief literature describes the need to construct a new world view and the importance of social support and rituals to come to terms with this loss (Hooyman & Kramer, 2006). Funerals can function as both a ritual and a means of bringing together supportive friends and family, but for animals whose grief is often overlooked or misunderstood, rituals may not be possible. This was not the case for Chester and other residents of Furever Farm where all animals have the right to grieve and farewell loved ones in their own unique ways and timeframes.

The dream of Furever Farm began in 2009 when a chance viewing of a documentary about the treatment of farmed animals transformed Darren and Hayley, two animal lovers, into activists and advocates determined to give animals a voice (Furever Farm, 2025). By 2016, Furever Farm was born with a simple vision and mission: to provide the dumped, abandoned and forgotten animals of the farming industry with love, care, devotion and commitment, while educating the community on the true nature of these sentient beings. Moving into the property at Rockleigh, South Australia, was only the start. There were fences to fix, shelters to build and water and feed troughs to install, while simultaneously taking calls for animals who needed new homes in this region where no farmed animal sanctuary had previously existed. Foundational residents included an orphaned lamb and ex-bobby calves but increasing demand saw goats, ex-racing horses, ducks, chickens, pigs—and the alpacas—find sanctuary at Furever Farm "Where the voiceless come to speak".

Ambrose and Chester arrived in 2020 when no longer wanted at their previous home. They had been together the whole of Chester's life, and Ambrose seemed to draw strength and confidence from her sociable and happy-go-lucky biological son. She arrived skittish and shy but was sustained by their loving and inseparable bond as she settled into this new environment. Chester remained unphased by anything that life threw at him during this exciting time, providing the perfect role model for his wary mother to learn that these humans could be trusted.

Like sheep, alpacas have been selectively bred to require annual shearing of the fleece so prized for luxury knitwear and yarns. For Ambrose and Chester who had not been shorn in over two years, relieving them of their thick and matted fleeces was a priority, even if it risked upsetting Ambrose's burgeoning confidence. There was relief all round when the expected struggle and vocal protests did not eventuate, and both emerged lighter and more comfortable. Unfortunately, the removal of her fleece laid bare Ambrose's emaciated frame, leaving everyone amazed at her tenacity to stand, let alone live, in such a weakened state.

By 2021, Chester's quirky personality brightened everyone's life, whether they were human, alpaca, cow, sheep or any of the other animal residents. Alpacas are naturally social, gentle and curious, and Chester met the brief to the full. His inquisitiveness knew no bounds, whether in his immediate vicinity or something in the distance to be approached at full speed and checked out. He also did not recognise personal space especially when placing his lips on his human friends and giving them big kisses.

Meanwhile, for a few short months, Ambrose regained a zest for life, forming bonds with the other alpacas and sheep, eating well and enjoying time with her cheeky son—but seemingly unable to gain weight on her skeletal body. While her body remained frail, Ambrose's eyes told a different story of happiness, love and joy in having what most farmed animal mothers never experience: a constant relationship with her son from beginning to end.

Alpacas have a vast range of emotions and ways of expressing them. While Ambrose focused on her daily nutrition-rich food, Chester and his alpaca friend Napoleon could sometimes be found practising odd facial expressions for reasons known only to themselves. Napoleon's rendition of alpaca sulking face—known as "droopy lips"—saw his bottom lip become droopy and fall away exposing his bottom teeth and gums, but this seemed to have no effect on Ambrose who would continue eating her special food. Napoleon and Chester had developed a strong friendship that would prove beneficial for Chester in the months to come.

In April 2021, Ambrose's 20-year-old frail body gave up, and she no longer had the energy to stand. Chester stood protectively by her side, talking to her constantly and leaning down to periodically kiss her tiny head. Only those he trusted were allowed to approach as his palpable concern and confusion culminated in unrestrained crying and urgency to continue kissing her head when she passed.

At Furever Farm, animals are grieved and buried in their own paddocks, with headstones as constant reminders that each life had been someone, not something. For smart, sensitive and emotional alpacas, the burial ritual is essential as they experience heart-rending devastation at the loss of a bonded parent or friend. When psychologist William Worden (2018) described the journey through grief in his "four tasks of grief" model (that is, acknowledge, process, adjust, and find an enduring connection), he was focused on human grief, but it could equally be applicable to the path that Chester appeared to follow at the death of his mother.

Acknowledgement of Ambrose's death came when seeing her lying in the earth before burial. This was followed by four days sitting atop his mother's burial site, as he appeared to slowly process the pain of his grief. Knowing she was gone did not stop him calling for Ambrose, the one who would normally have helped him adjust to this life-changing event. But Chester had support in the humans who allowed him to mourn for as long as he needed, and alpaca friend Napoleon who stood behind him, only wandering off to feed before returning to stand vigil with the grieving son. In his own time, Chester returned to the paddocks to wander with his remaining family group as he slowly adjusted to living without his mother by his side. However, an enduring connection remained during the occasional visits to his mother's resting place for reasons known only to himself. Chester and Napoleon developed a stronger bond than ever, contributing to the gradual reinstatement of Chester's naturally happy disposition.

Animal residents at Furever Farm have the opportunity to live out their full life free from abuse, neglect and slaughter. But many enter the sanctuary already damaged from cruelties previously inflicted, and while their psyche may be mended, the physical damages may shorten their longevity. Napoleon entered Furever Farm at 14 years of age, having been discarded when his alpaca stubbornness and non-compliance with human plans made him too much trouble to bother with any longer. Napoleon could not trust, crying, squealing and spitting as his overgrown fleece revealed a beautiful, collaged body and unique face markings with half a moustache. Napoleon eventually learned to trust, even granting the humans the privilege of touching him and responding to his delicate whines to be hand fed. But in 2024, Napoleon bowed his head, went to sleep and passed away peacefully.

Once again, Chester sat atop a grave for days as he acknowledged and adjusted to the loss of the faithful companion who would normally be standing vigil beside him. Anecdotally, those living with alpacas report excellent long-term memory in these intelligent sensitive animals, making it likely that Chester's past experiences of death and burial triggered memories and coping mechanisms as he worked through this recent loss. Increasingly, the conversation among animal behaviourists has moved from "Do animals grieve?" to "How do animals grieve?" (Pierce, 2018), adding urgency to the question of what humans can do to acknowledge and support animal grief. Chester was fortunate to be living at Furever Farm where each resident is given the opportunity to grieve in their own unique way.

Body language and vocalisations

Once a person knows what to look for, and listen to, alpacas and llamas do not hold back in expressing their feelings. Body language and vocal communications in particular play a significant role in exchange of information between members of the social group and provide strong indicators of what is going on in the animal's thoughts and feelings (la Lama & Villarroel, 2023), as Napoleon's sulking

Figure 12.1 Chester stands protectively by his mother Ambrose's side. (Image supplied by Furever Farm and reproduced with their permission.)

face demonstrated. Llama ears are taller than alpaca ears, but both make good use of them when conveying important feelings (Barnes, 2022a). Ears up and forward facing can suggest curiosity, while ears drooping to one side show a relaxed and calm alpaca or llama. Once the ears start pointing backwards, levels of displease increase, culminating in ears pinned all the way back and head up, suggesting that the alpaca or llama is either very displeased or uncomfortable, or both.

Spitting also plays a part in the llama and alpaca's emotional repertoire, often in response to feeling threatened by an invasion of personal space, while vocalisations further express their feelings, whether happy, sad or grumpy. Alpacas can be more vocal than llamas, but vocalisations in both provide a conduit to their emotional states, social relationships and health issues (Barnes, 2022b). Humming expresses moods and feelings, and depending on tone and volume, every hum will have a different meaning. There are separation hums, distress hums, displeasure hums, concerned hums and relaxation and contentment hums. Friends may hum together as they nuzzle each other, and a mother will emit a low-pitched hum as she bonds with her cria who soon learns to reciprocate. Listening to them take turns in their conversation reveals a deep connection between mother and infant.

Alarm is conveyed through staccato high-pitched squeaky braying, but clicking is a little more complex. There can be a mild back off meaning, or a hello, or even a mother's concern about her cria—each needs interpreting through the lens of the animal's body language. Groaning and moaning are self-explanatory, while grumbling suggests displeasure and a warning to back-off. A deep-throated growl reveals a very grumpy alpaca or llama. Orgling, which sounds a bit like a snorting trumpet player, can be heard when a male chases his chosen lady or has already started mating with her. Sometimes, the lady will express displeasure by snorting air through her mouth to say she is not interested and to leave immediately. Fearful or angry emotive states and trauma—such as being tied down during shearing—prompt screams, squealing or screeching.

Each alpaca and llama bring their own unique nuances to these vocalisations, but combining the situation's context and body language provides a clear representation of how these animals with a huge emotional capacity are feeling at any one time.

Unlikely friends at Magzalea Farm and Sanctuary, USA

In New Hampshire, USA, a very special farmed animal sanctuary sits on almost 40.5 hectares (100 acres) of rural land (Magzalea Farm and Sanctuary, n.d.). Like many farmed animal sanctuaries, from the moment Magzalea Farm and Sanctuary opened its doors, the animals just kept coming, forcing them to quickly acquire some extra land. From the early days, this family-run sanctuary chose to focus on the long-term care of senior, orphaned, disabled and other animals with special needs, and the demand never slowed. These animals, affectionately called "Magzalea Misfits" and defined as "A unique and special soul who is too awesome to live just anywhere", are daily proof that many physical and psychological scars can be eased in a caring and respectful environment. Inevitably, the scars can never be totally erased for many of the animals, especially those escaping human abuse and exploitation, but Magzalea offers every resident the right to live and die in a safe, loving forever home.

For over a decade, a group of alpacas have wandered the grounds of Magzalea Farm and Sanctuary. Each one has a unique and quirky personality which, once comfortable in their new home, is openly displayed as they make their preferences known. Protectiveness features in the group, ranging from Sammy who will sneak up and bite or kick those administering vaccinations or medications to his friends, to the elderly Honey who takes the matriarch role seriously with her distinctive chirp or scream to alert her friends to anything unusual. Others are content to be left alone to roll in clean bedding, enjoy a good belly wash with the hose or just rummage around to see what they can find.

But these alpacas also share a very special trait—they make it clear how they want to relate to other species, whether that be humans, sheep, an emu, a rabbit or a pig. For some of the alpacas, their idea of a good life has diverged in unpredictable ways from the species-typical norms, allowing them to find enjoyment in unexpected relationships. In a setting bearing elements of Donaldson and Kymlicka's (2015) farmed animal sanctuaries as "intentional communities", the alpacas have clearly shown their preferences on how to live in an interspecies community.

Tia, one of the senior alpacas, has found a best friend in Greta, the emu who seems to identify as an alpaca. Emus, large flightless birds native to Australia and farmed for the oil extracted from their fat, typically live in pairs and travel in mobs. Greta does everything with Tia—eating, sharing the wading pool, lounging around in the sun together—in fact, Greta is the only one that Tia allows to take a sneak bite of her precious food. Emus love to swim, and Greta can regularly be seen lining up with the alpacas for a dip in the pool or fun under the sprinklers.

Then, there is the gentle, happy P Man who loves his sheep family and can often be found in the wooded area lying around with the sheep, while Zeb the elderly sheep reciprocates by mingling with the alpacas. Charlie, the one llama at Magzalea, was rescued from a kill pen and took on a big brother protective role to both the smaller alpacas and the sheep family.

When Poppy, the first donkey to join the sanctuary arrived, she was welcomed by the alpaca group until she got over her fear of humans. And a more unlikely friend to the alpacas and all the other residents is Nomie (aka Nomad) the rabbit who found his way to the sanctuary and settled right in. Nomie does not recognise boundaries with any of the sanctuary animals, but always makes a point of visiting the alpacas and goats to help clean up any hay they may have dropped. There can be logistical difficulties in maintaining a shared community, but at Magzalea, species discrimination does not exist for the residents.

Something very special happened a decade ago that demonstrated the power of interspecies connection. An older pig named Petunia hated to be touched. Although not fearful of humans, she would nevertheless squeal and snap if anyone—human or nonhuman animal—tried to lay a hand on her. That is, until she met Bubblin', a blind alpaca cria who joined the sanctuary with her mother. A connection was made and best friend Bubblin' became the first and only individual to touch Petunia without a squeal or flinch.

Magzalea's intentional focus on the so-called misfits has revealed possibilities for animals and humans alike in ways that no one could have predicted when one family chose to share their lives and their home with these unique sentient creatures.

ETHICAL IMPLICATIONS

Alpaca and llama welfare has received less research attention compared to other farmed animals such as cattle and pigs. Many alpacas and llamas fall in that grey area between large-scale industrial farming systems and small landholders, which has further compounded the difficulties in standardising welfare legislation at a global—or even national—level.

Figure 12.2 Ditto (front) and the alpacas at Magzalea Farm and Sanctuary all have unique personalities and choose their own friends. (Image supplied by Magzalea Farm and Sanctuary and reproduced with their permission.)

Despite relatively few physiological changes brought about by domestication compared to farmed animals selectively bred to grow bigger and faster or lay more eggs or produce more milk, alpacas and llamas still show signs of what has been labelled the "domestication syndrome" (du Toit, 2020, p. 303). These include increased docility and trust, reduced aggression, changes in brain size and skull shape and specific selection of coat colour, texture and density irrespective of potential welfare concerns. Cumulatively, these changes have primed alpacas and llamas to be providers of both emotional and physical support in their roles as companion animals, therapy animals, tourism and entertainment animals, small farm animals, factory farmed animals and research animals. Within each role, some research exists in a positive trend towards a deeper understanding of these sentient creatures.

Alpaca and llama meat provides one of the main protein sources for Andean populations, but has not yet secured a strong representation in other countries. This may change as interest and research outside of South America increases and other countries investigate this high protein meat source from an animal that has proven to contribute to a more sustainable environment. This escalates the urgency to gain a greater understanding of these two unique animals before demand for both their meat and fleeces continue the relentless progression towards industrialised farming systems with associated negative welfare implications. Of course, given the alpaca and llama's "cute factor" and their role as companion animals and therapy animals in western countries, it may prove difficult for some people to reconcile the cognitive dissonance associated with eating the close relatives of their companion or therapist.

REFERENCES

Australian Alpaca Association (2020). *Introduction to alpacas for RSPCA officers.* https://alpaca.asn.au/animal-welfare/#rspcaoverviewaugust2020

Balcarcel, A.M., Bastiaans, D., & Orliac, M.J. (2022). Endocranial casts of *Camelops hesternus* and *Palaeolama sp.*: New insights into the recent history of the Camelid brain. *Brain Behaviour & Evolution, 98*(2), 107–120. https://doi.org/10.1159/000528762

Barnes, A. (2022a). Understanding Camelid body language Part 1: Ears. The Open Sanctuary Project. https://opensanctuary.org/understanding-camelid-body-language-part-1-ears/

Barnes, A. (2022b). What's the word?: A glossary of camelid vocalizations and sounds. The Open Sanctuary Project. https://opensanctuary.org/whats-the-word-a-glossary-of-camelid-vocalizations-and-sounds/

Bekoff, M. (2007). *The emotional lives of animals: A leading scientist explores animal joy, sorrow, and empathy--and why they matter.* New World Library.

Bentham, J. (1789). *An introduction to the principles of morals and legislation.* Clarendon Press.

Donaldson, S., & Kymlicka, W. (2015). Farmed animal sanctuaries: The heart of the movement? A socio-political perspective. *Politics and Animals, 1*(1), 50–74.

Du Toit, J. (2020). The ethics of domestication. In B. Fisher (Ed.), The Routledge handbook of animal ethics (pp. 302–315). Routledge.

Duncanson, G.R. (2023). *Veterinary treatment of llamas and alpacas.* CAB International.

Fine, A.H. & Griffin, T.C. (2022). Protecting animal welfare in animal-assisted intervention: Our ethical obligation. *Seminars in Speech and Language, 43*(1), 8–23. https://doi.org/10.1055/s-0041-1742099

Furever Farm (2025). *Our vision.* https://fureverfarm.org.au/our-vision

Gunsser, I. (2013). *Animal welfare problems in alpacas and llamas in Europe.* 64th EAAP Annual meeting, 25–30 August, 2013, Nantes, France Symposium on South American Camelids and other Fibre Animals. https://docs.eaap.org/2013/S43_11.pdf

Hall, H. (2019). Meet the alpacas that are helping researchers who study autism, Alzheimer's and cancer. *Vanderbilt University Research News.* https://news.vanderbilt.edu/2019/08/13/meet-the-alpacas-that-could-cure-autism-alzheimers-and-cancer/

Hill, E. (2024, October 21). King Charles caught off guard by sneezing alpaca in unexpected royal tour moment. *People.* https://people.com/king-charles-caught-off-guard-sneezing-alpaca-australia-tour-8731490

Hooyman, N.R. & Kramer, B.J. (2006). *Living through loss. Interventions across the life span.* Columbia University Press.

Jost, S.M., Knoll, A., Lühken, G., Drögemüller, C., & Zanolari, P. (2020). Prevalence of coat colour traits and congenital disorders of South American camelids in Austria, Germany and Switzerland. *Acta veterinaria Scandinavica, 62*(1), 56. https://doi.org/10.1186/s13028-020-00554-y

la Lama, G.C.M-d. & Villarroel, M. (2023). Behavioural biology of South American domestic camelids: An overview from a welfare perspective. *Small Ruminant Research, 220*. https://doi.org/10.1016/j.smallrumres.2023.106918

Magzalea Farm and Sanctuary (n.d.). *About*. https://www.magzaleasanctuary.org/about

Mathews, P., Doyle, E., Doughty, A.K., Brown, W.Y., & Barwick, J. (2024) Using alpacas as guardian animals in Australia: A survey of sheep producers. *Animal Production Science, 64*, AN24093. https://doi.org/10.1071/AN24093

Montelli, S., Stelletta, C., Ruiz, J.A., Ballarin, C., Cozzi, B., & Peruffo A. (2019). Mapping of the early intrauterine morphogenesis in the Alpaca (Vicugna pacos): External features and development of the Cephalic Vesicle in comparison with the Progressive Carnegie Scale. *The Anatomical Record, 302*(7), 1226–1237. https://doi.org/10.1002/ar.23980.

PETA (2024). *Groundbreaking undercover investigation: Crying, vomiting alpacas tied down, cut up for sweaters and scarves*. https://investigations.peta.org/alpaca-wool-abuse/#action?en_txn7=blog::llama%20versus%20alpaca

PETA (2025). *'Llama' tell you something—Alpacas and llamas are very different*. https://www.peta.org/features/alpaca-and-llama-differences/

Pierce, J. (2018, August 24). Do animals experience grief? *Smithsonian Magazine*. https://www.smithsonianmag.com/science-nature/do-animals-experience-grief-180970124/#:~:text=A%20growing%20body%20of%20scientific,for%20or%20ritualize%20their%20dead.

Understanding Animal Research (2025). *Llama*. https://www.understandinganimalresearch.org.uk/using-animals-in-scientific-research/animal-research-species/llama

Vilá, B. & Arzamendia, Y. (2022). South American Camelids: Their values and contributions to people. *Sustainable Sciene, 17*, 707–724. https://doi.org/10.1007/s11625-020-00874-y

Worden, J.W. (2018). *Grief counseling and grief therapy: A handbook for the mental health practitioner* (5th ed.). Springer Publishing Company. https://doi.org/10.1891/9780826134752

13

Camels: Majestic and resilient wanderers

INTRODUCTION

As the planet experiences signs of global warming, one animal's adaptability has drawn increased attention. Dromedary camels are one of the few large mammals that can thrive in harsh desert environments. Their physiological and morphological adaptations for water conservation and thermoregulation allow them to continue functioning and thriving on low-quality feed, making them a potentially sustainable source of meat and milk for humanity in these changing global conditions (Abdullah & Sheikh, 2024). With their stoic capacity to bear pain, they can continue working without showing signs of suffering, and their long acclimatisation to desert environments strengthens their ability to walk for up to a week without food or water and lose up to a quarter of their body weight without their vital body functions closing down (Previti et al., 2016). Already the fifth largest contributor to global milk production, demand for camel milk is estimated to increase manifold in areas where summers grow increasingly arid.

But camels are so much more than resilient commodities to service human needs. They are sentient animals whose welfare has been largely overlooked in research programmes and legislation, and whose treatment continues to be heavily influenced by local traditions. In areas such as Saudi Arabia, Africa and Asia, the welfare of camels is strongly embedded in inherited cultural practices that determine all aspects of their treatment from breeding to rearing to medical interventions. In Australia where feral camels survive and thrive, especially in drought years, farmers shoot hundreds of these "hardy pests" on a regular basis.

There are two distinct species of camel—the Bactrian camel and the dromedary—with some crucial physiological differences (Choudhary, 2021). The Bactrian camel, mostly found in the colder climates of Central Asia, has two humps which flop over when depleted by starvation. The dromedary, commonly found in the warmer climates of Middle East and North Africa, has one hump with elastic skin that allows it to gradually diminish in size rather than

DOI: 10.1201/9781003518853-14

collapse under starvation conditions. Most of the research and literature to date has focused on the dromedary in its role as a highly specialised desert animal. In this chapter, the term camel is used to signify the dromedary unless otherwise specified.

SOME BACKGROUND HISTORY

Evidence of the family *Camelidae* appeared around 40 million years ago in North America. Molecular studies indicate a split into Old and New World camelids somewhere between 11 and 25 million years ago, with the old-world camelids crossing the Bering Strait land bridge during a glacial period in the Pleistocene Epoch to the eastern hemisphere and further splitting into two species: *Camelus bactrianus* (two-humped camels) and *Camelus dromedarius* (one-humped camels) (Amandykova et al., 2023). There was also an undomesticated two-humped camel, *Camelus ferus*, which separated from domestic two-humped camels somewhere between 0.6 and 1.8 million years ago. New World camelids (guanaco, vicuña, llama and alpaca) headed south across the Central American land bridge around the same time, and both new and old camelids continued evolving in harsh, but different environments, with sparse, poor-quality forage for some parts of the year. Old world camel domestication was established by around 4,500–5,000 years ago in South-East Arabia and South-West Central Asia (Masebo et al., 2023).

One-humped camels, the dromedaries, largely settled in the southern hotter desert and semi-desert areas of the Middle East and North Africa, and the two-humped camels in the northern colder areas of Asia. The realisation that camels could survive and continue providing meat, milk, wool, hair, dung, transportation, military support during human wars and entertainment under the most adverse environmental conditions provided strong incentive for domestication, thus playing a significant role in Old World ancient nomadic civilisations' prosperity.

The camel became a symbol of individual and communal honour, often slaughtered for grand feasts such as weddings or funerals where the camel would be buried with human corpses (Alwahaibi et al., 2023). Those camel-herding Bedouin who raised camels in desert areas were traditionally regarded as the noble Bedouin, in contrast to the "common tribes" who raised sheep and goats on agricultural lands. Racing camels occupied the top of the hierarchy, kept in enclosures with just a few other camels and fed exclusively on the highest quality foods.

Retention of the same set of chromosomes meant dromedary and the Bactrian camels could be crossed to produce fertile offspring with muscular body constitution and strong endurance for transporting goods and for military use (Amandykova et al., 2023). Widespread hybridisation in Türkiye produced the large animals favoured for annual camel fights and in Kazakhstan to create more productive camels to withstand the harsh climate of Central Asia. Hybrid wool production was deemed greater than pure dromedary breeds and hybrid milk production greater than Bactrian camels.

In 2021, the African nations of Chad, Somalia and Sudan had the largest camel populations, with 9.4, 7.4 and 4.9 million camels, respectively, largely in a subsistence and extensive system of migratory pastoralists in arid and semi-arid regions (Masebo et al., 2023). In Central Asia, Kazakhstan's camel population of Bactrians, dromedaries and hybrids had reached 2.6 million by 2022, making camel breeding economically profitable for the provision of wool, meat and milk (Amandykova et al., 2023). Some reports estimate that the world camel population could increase from the estimated 41.1 million head in 2020 to around 60 million within 25 years as climate change makes this versatile animal more and more valuable.

CAMELS IN CONTEMPORARY SOCIETY

Any resilient domesticated animal capable of fulfilling multiple human needs is at risk of compromised welfare and the camel, especially the one-humped dromedary, is no exception. Dromedaries have social meaning and economic significance for human livelihoods in the arid and semi-arid environments of Africa and Asia (Padalino & Menchetti, 2024). They provide a sustainable income through the sale of camel milk, meat, hides and wool, while concurrently shaping cultural practices and traditions and strengthening social bonds in these communities. Camel herding across wide areas enables communities to access remote markets and the camels to experience a lifestyle more in tune with their natural behaviours. However, with droughts and unpredictable rainfall cycles extending across time and geographic location, even the hardy dromedary can face challenges. Hunger and thirst during unusually long dry seasons threaten their welfare and lives, especially in remote nomadic areas with no veterinary service and no essential drugs to counter potential disease, parasites, injury and distress.

The increase in industrial farm systems however introduced a different set of welfare issues for these animals who, under natural conditions, liked to wander across large areas, "ambulatory grazing" as they walked (Menchetti et al., 2021, p. 532). More comfortable and familiar with small stable groups of around nine members, the reduced space allowances of intensive systems negatively impacted both individual and social wellbeing. Placing camels in overcrowded pens without the ability to walk freely where and with whom they wished risked the formation of unstable social relationships, increased levels of aggressive behaviours, stereotypies, compromised feeding and drinking behaviours and environmental degradation.

Camel intensive farming systems were set in motion following the development of mechanised camel milking machines in the early twentieth century (Abdullah & Sheikh, 2024). By 1970, these machines were adopted in the former Soviet Union, eventually culminating in the world's first large-scale camel automatic milking parlour in Dubai in 2002. Additional machine-milking countries, including other parts of the United Arabs Emirates, Saudi Arabia, Tunisia and small-scale farms in Australia, Europe and the USA, followed this trend. The largest camel dairy farm in the United Arabs Emirates eventually housed more than 10,000 camels, including dairy cows selectively bred to produce higher milk yields and fattening units for male camels to be slaughtered for meat (Ahearn & Chatty, 2024).

Large-scale intensive animal agriculture prevents animals expressing all their natural behavioural needs and has become synonymous with the development of stereotypies among a range of confined animals. Stereotypies are functionless abnormal locomotor or oral behaviours triggered by a lack of control over the environment, social isolation, frustration, threat, fear or lack of stimulation (Padalino et al., 2014). They provide a form of self-soothing pleasure when endorphins are released, thus reinforcing the behaviour and helping an animal cope with captivity or confinement stressors. Male camels are particularly vulnerable when kept in isolation between programmed mating, deprived of social interactions in herds and unable to spend the day walking and grazing. When assessing husbandry among male dromedary camels, animal behaviour and welfare researcher Barbara Padalino and colleagues (2014) reported both oral and locomotor stereotypies in isolated male camels, including self-biting, bar-mouthing, headshaking and pacing in a circle. Breeding season heightened these behaviours when males were confined or tethered to counter aggression towards other camels and humans.

This aggression was capitalised on during the annual camel wrestling festivals in Türkiye. In a tradition believed to date back hundreds of years among competing nomadic groups, the games take place during the camels' mating season when the males are particularly aggressive, especially if barred from mating. A female camel on heat is located nearby and a winner declared when one male retreats, screams or falls. Animal fights are banned in Türkiye, but camel wrestling is positioned as a traditional folkloric show despite animal rights groups calling for the global ban of this exploitative use of animals.

Tourism

Similar to bobby calves, most male camels born into the dairy industry do not fare well. Unwanted bulls from dairying may be sold at camel markets for meat or forced into the tourist or entertainment industry, once their will has been painfully broken (Narayanan, 2021). Traditional training practices to ensure tourist safety require dominating and breaking the camel's will, often by cruelly knocking them to the ground and tying the animal's legs. They may be socially isolated, starved, have ropes and sticks pushed through their sensitive noses, hobbled, muzzled and forced to continue working when injured or exhausted—all so tourists can ride on their back and be photographed. Between 2019 and 2024, People for the Ethical Treatment of Animals (PETA) Asia (2025) publicised images of horrific abuse in an Egyptian camel market where camels were being sold and exploited for rides and photo opportunities at the nearby pyramids. Animals were beaten into submission, exploited until their exhausted bodies gave out and then returned to the market to be sold as meat. With legs tied to prevent escape and bloody fly-infested wounds causing relentless pain, these patient sentient animals suffered unimaginable physical and psychological pain before their slaughter.

Camel rides are ubiquitous in many other countries and regions, including United Emirates Republic, Jordan, Morocco, the USA, Canary Islands and Australia. Piebald camels are particularly valued in the tourist industry, including the Canary Islands and the USA where they are actively bred for tourism. Their aesthetic appeal translates into a higher monetary value compared to solid-coloured camels, thus increasing demand (Volpato et al., 2017).

Australia's use of wild camels for riding tours raises many ethical issues, including confinement, restraint and questionable training methods for animals who were born free. Around 20,000 camels arrived in Australia in the 1840s to assist with exploration and transportation across desert and arid regions (Farm Transparency Project, 2020). The arrival of rail and motor transport in the 1930s eliminated their economic practicality, resulting in mass abandonments. By 2008, the hardiness and adaptability of these animals saw their population reach an estimated 800,000, and by the 2020s, an estimated 1.2 million. This hardiness, once so appealing when camels served human needs, became a liability as they thrived in the arid outback to become the largest wild population in the world, simultaneously achieving invasive pest status. With this status came culling, primarily through shooting from helicopters or rounding them up and trucking them to slaughterhouses, live export, dairy farms or tourism. The forced seizure, transport, handling and confinement of these wild free-ranging animals caused extreme distress with long-term physical and psychological implications for those camels not immediately slaughtered.

Training of camels destined for the tourist trade in Australia necessitates restraint, whether by tethers or hobbles tying two legs together, both of which are distressing for a wild animal in an unfamiliar and threatening environment (RSPCA, 2024). Some operators use nose pegs through the soft tissue of the nose or upper lip to assist handling but causing ongoing pain. Most tours operate in the middle and northern outback regions of Australia where temperatures regularly reach over 40 degrees Celsius (104 degrees Fahrenheit), and even the stoic camel needs access to shade and regular rest schedules. This does not always happen, and with only a non-mandatory Code of Practice for camel welfare, reports indicate some camels in tourism continue to suffer. As non-native animals with pest status, feral camels are not subject to the same regulations and protections as native wildlife, raising multiple ethical and welfare issues around their capture, breaking-in process, housing, ongoing treatment and eventual slaughter.

Part of Australia's camel tourism also focused on camel racing, including the famous 1988 Great Australian Camel Race spanning 3,300 kilometres from Uluru, the centre of Australia, to the beaches of the Gold Coast in Queensland. While camel racing in Australia draws tourists each July, it lacks the high stakes of camel racing in the Middle East regions.

Camel racing and camel beauty pageants

With the diminishing need to herd camels as a means of transportation in Bedouin communities, the focus shifted to racing camels and the vast prize money at festivals and other events sponsored by wealthy sheikhs and organisations (Alwahaibi et al., 2023). Proven race winners and their progeny sold for massive amounts of money, reinforced by the allure of ever-increasing prize winnings. Dating back to the pre-Islamic era when Arab tribes would gather to show off their camels' strength, camel racing instead transitioned to a competitive investment for earnings from oil and gas. Camels start their racing career at two to three years of age, and assuming that they are deemed worthy to continue racing, their career can last until they are eight or nine years. As with many valuable animals used for high-stakes human entertainment and sport, they can be attentively nurtured but still at risk of some welfare concerns when preparing for races. Racing camels can experience lameness, forced weight-loss resulting in ketosis or laminitis and abnormal behaviours from a change in feed during the racing season (Al Jassim, 2024).

Compounding the commodification of camels in a competitive environment are the camel beauty contests that draw in thousands of camels from across the Arabian Peninsula to compete for tens of millions of dollars in prize money. Make-up, tail fluffing, coat oiling and glossing of the camel's long eyelashes, which in their natural environment provide protection from sandstorms, are the norm as camels are modified to comply with human perceptions of beauty ("Of Botox and camel beauty contests", 2018). By the twenty-first century, camels were being subjected to more invasive procedures despite cosmetic surgery being illegal. In 2018, twelve camels were disqualified from one of the most prestigious festivals following the use of Botox injections to plump up their lips, noses and jaws. Three years later, the use of sophisticated devices to detect tampering, including X-ray machines and sonar devices, resulted in 40 camels being disqualified and the organisers' reiteration that all acts of tampering and deception in the beautification of camels would be rigorously prevented ("Botox injected into camels", 2021). Citing examples such as Botox injections to relax muscles around the camels' faces, collagen fillers to make the camels' lips and noses bigger, hormones to boost muscle growth and rubber bands to make body parts bigger by restricting the blood flow, the emphasis was on catching and penalising these offenders rather than working to end the exploitive and cruel beauty pageants imposed on a sentient animal.

FEELINGS AND EMOTIONS

Prior to recognition of the camel's environmental adaptability and economic significance in a changing global climate, the camel had attracted less research than other domesticated animals. Like many other mammals, their brain anatomy includes the neocortex's intense gyrification (folding), which is crucial for information processing, consciousness, memory and motor function and the hippocampus, associated with forming and retrieving memories, spatial navigation and emotional responses (Al Aiyan et al., 2024). Their overall brain weighs less than horses and water buffaloes, but more than many cattle species, and importantly, it includes a unique cooling mechanism that protects brain tissue and allows camels to survive in high temperatures.

As camels faced an increasing number of lifestyle adaptations driven by climate change and human demand, temperament played a crucial role in adapting to and surviving novel situations. In mammals, the monoamine oxidase A (MAOA) gene has been associated with personality and feelings of anxiety, fear, excessive worrying and tension. The MAOA gene, located on the X chromosome in dromedary camels, codes for an enzyme instrumental in the breakdown of neurotransmitters such as serotonin, dopamine and norepinephrine, thus maintaining a constant level of these chemicals in the brain for mood regulation (Ramadan et al., 2018). Knowledge of camel genome assembly may lag behind other domesticated species, but the camel's multi-functional purposes have seen an increased interest in assisted selection of desirable "improvements" to the camel personality and mood to match their designated use. This acknowledgement of the presence of neural mechanisms influencing personality and mood validates camels as sentient beings but also leaves them open to human manipulations to meet the demands of intensive production systems, racing events, tourism and entertainment.

Given the ethical and welfare issues already identified around intensive animal agriculture, there is an urgency not to follow the same path with animals who have received less attention than the more traditional farmed animals. In the 1980s, arguments arose as to whether genetically altering farmed animals to make them more accepting of their intensive farming environment could be an answer to welfare issues. "Disenhancement" favoured the creation of domestic animals that were more suited to industrialised farming by reducing their capacity to experience negative affective states. In a paradoxical welfare justification, these arguments suggested that selectively applying genetic manipulations to impoverish an animal's mental life would eliminate their capacity to suffer and solve the immediate animal welfare concerns associated with this system (Monsó & Hintze, 2023).

However, opponents of this approach acknowledged that while disenhancement could improve farmed animals' current situation and turn a miserable life into a manageable one, it could also impede farmed animals from experiencing a good life. Diminishing an animal's mental life so that they had reduced capacity for negative emotional states introduced the risk of also diminishing their individuality and capacity for social relations, emotional contagion and caring (Monsó & Hintze, 2023). Positioning disenhancement as a solution to animal suffering instead avoided the bigger picture of the oppressive environment that created this suffering. With more and more dairy camels at risk of being born into industrialised farming systems, disenhancement would remove the very traits that make them unique sentient individuals.

When two young camels from dairy farms arrived at Strong Hearts Farm Sanctuary in Victoria, Australia, no one could have predicted the bond that would form between these sensitive creatures, and the emotional rollercoaster that would follow.

The unique personalities of Carter, Thyra, Kira, Rue and Sasha—Strong Hearts Farm Sanctuary, Australia

Underweight and forlorn, Thyra arrived at Strong Hearts Farm Sanctuary when her mother, a dairy camel, became too sick to look after her. According to a 2020 survey among Australian commercial dairy camel producers, most of their dromedary camels were sourced from the feral camel population in Central Australia (Fallon et al., 2020). This meant that many arrived traumatised from being caught in the wild and transported long distances. Even though dromedary camels are less efficient milk producers compared to cows, camel milk is an emerging market in Australia and sells for considerably higher prices than other animal-derived milk. Australian dromedary camels produce around five litres a day during their 40- to 78-week lactation period.

Thyra's health and weight fluctuated over the following year, and despite medications, fluids, tests, vet consultations, special feeding and tender care, Thyra passed away. Carter, who had also arrived from a dairy farm, struggled to cope with Thyra's death and his lone-camel status. With plummeting emotions, Carter continued to call out for Thyra and could be found staring at his reflection in a plastic storm cover on the stables. As herd animals, camels form small stable groups, and the goats and cows were no substitute for his bonded camel friend, Thyra. Grief and social isolation threatened Carter's psychological wellbeing.

Camels have largely been omitted from the debates around genetically modifying farmed animals to eliminate negative affective states associated with the pain, fear, anger, frustration, sadness, boredom and grief endemic in commercial farm systems. With less research into the camel's genome assembly and welfare indicators, understanding their emotional capacity has been hindered by the perceived stoicism of these majestic animals masking both their positive and negative affective states, especially during transport, confinement and slaughter. Bedouin cameleers who lived closely with their camels learned to recognise and understand a wide range of emotions among group members, as do those who rescue and rehabilitate camels in farmed animal sanctuaries where the camels' personalities emerge with safety.

Carter's suffering at the loss of Thyra and the meaningful bond they shared was painful to watch, but it also meant that he retained the capacity to experience future positive affective states once the grief subsided. Genetic manipulation of an animal's capacity to suffer would allow them to adjust to psychologically isolated lives in industrial settings, but also prevent the formation of meaningful bonds, including maternal bonds, throughout their short lives. When Carter met Kira, another camel from the dairy industry, the return of his positive affective state demonstrated the importance of retaining emotional capacity to live a good life.

Kira came from the same dairy farm as Carter and was meant to join the sanctuary alone—until the team went to collect her and noticed two sickly young camels. These young camels had been taken off their mother early when she became unproductive through excessive weight loss. Stunted and in desperate need of some good nutrition and care, it was impossible to leave the bonded pair behind. The arrival of Kira, Rue and Sasha had an immediate effect on Carter, as described by the Strong Hearts team:

> If Carter's jaw could have dropped in shock when his new friends arrived, it would have! His whole mood and demeanour changed overnight, and he is overjoyed to have camel companions again. The group are already inseparable, and all three girls are super curious and inquisitive. This truly highlights the importance of herd animals having companions of their own species.
>
> *(Reproduced with permission from Strong Hearts Farm Sanctuary, Australia)*

While hypothetical discussions of disenhancement through genetic manipulations may never eventuate among the feral camels who largely service the dairy industry in Australia, the risk nevertheless remains as camel dairy farms increase in size and breeding in captivity increases.

Before long, a new camel entered Carter's life when Zeebah, whose mother had been culled as a "pest species", arrived at Strong Hearts. Carter was besotted and often caught gazing at her, mesmerised. Carter's reaction to this beautiful camel

Figure 13.1 Carter's mood and demeanour changed with the arrival of Kira, Rue and Sasha. (Image supplied by Strong Hearts Farm Sanctuary and reproduced with their permission.)

Figure 13.2 Best friends, Carter and Kira. (Image supplied by Strong Hearts Farm Sanctuary and reproduced with their permission.)

and his newfound happiness after grief is testament to the importance of retaining both negative and positive affective states in an animal's life.

In addition to a cooling system, the camel's brain has other unique features, attributed to their need to manage the pain of living in harsh climatic conditions and exposure to painful stimuli such as chewing thorny plants and cacti. Researchers Qasim A. El-Dwairi and colleagues (2023) proposed that painful sensations in the facial area may affect the development of neurons that receive information about touch, pain and temperature in these regions, making the camel appear very stoic and expressing only subtle signs of fear, pain and distress. There is also some evidence that neurons in the superior colliculus of the camel brain are associated with a pain-inhibiting opioid pathway, which provides an additional adaptive trait to living in extreme temperatures and environments (Benaissa & Iglesias Pastrana, 2024). The camel is not immune to pain though, as nociceptive markers have been identified in various regions of the spinal cord, including the lumbar dorsal root ganglion and the soft tissue of the nasal septum in dromedaries, making the insertion of nose sticks and rings for training and control particularly problematic. No studies have been conducted on the association of emotional pain and physical pain in the stoic camel, an area that is of importance as commodification of camels intensifies. However, the identification of stereotypies in confined camels does indicate psychological pain. Similarly, PETA Asia's images of camels screaming in pain at markets and when crude methods of slaughter are applied suggest that physical pain is a stressor for these sentient animals, irrespective of their neural adaptations.

As with research into the camel's genome assembly, studies on positive welfare indicators lag behind other domesticated animals and have been hindered by the perceived stoicism of these majestic animals masking any emotional expression. Anecdotally, Bedouin cameleers believed in their camel's ability to experience positive and negative emotions appropriate to a situation, arousing a reciprocal emotional response in the cameleer and making the camel's status equal to humans in Bedouin culture (Alwahaibi et al., 2023). Bedouin cameleers also believed that their camels could analyse the circumstances and select a line of behaviour based on the expectations of their owner, a response triggered by their close contact and continuous interactions with humans. Among the Khusman Bedouin in the eastern desert of Egypt, camels received names that described their special characteristics, thus endowing the camel with both identity and emotional capacity. Unfortunately, this bond between cameleer and camel, which recognises the individuality of each camel, is not replicated in the growing number of intensive farm settings, where these feeling animals become faceless commodities.

Away from human influence, camels are social animals whose psychological wellbeing can be negatively impacted by social isolation. Capitalising on this need to be with other camels, a cost-effective programme to control the feral camel populations in Australia was trialled between 2008 and 2010 (Spencer et al., 2015). After confirming the homogenous genetic population structure across Australia and the group fission–fusion system, whereby social units were transient and dynamic rather than bound by kinship, a single tagged camel was sent in to betray the location of hard-to-reach groups. Controversially named the "Judas method", the lone camel was compelled to seek out and "betray" the location of a group of feral camels who could then be culled. Using a lonely sentient animal's need for social connection as a weapon has been employed among a number of so-called pest species, including feral goats, pigs and donkeys.

Camels connect and communicate their emotional and physical state in many ways and with many sounds, from moans to loud bellows, grunts to loud roars, humming between mothers and newborns, blowing in each other's faces as a friendly greeting and different positioning of their head, neck, ears and tail (Nature, 2020). A roar or a hiss when confronted by a predator can give way to a soft grunt of contentment during safe times of grooming or feeding. Not surprisingly, when faced with stressful conditions, camels tend to vocalise more, an observation reported when comparing increased vocalisations among camels kept in direct sun during extreme heat conditions and those with access to shaded areas (Papageorgiou & Simitzis, 2022). Footage from PETA's investigation of camel markets and slaughterhouses leaves no doubt that camels vocalise when pain reaches an unbearable level, highlighting that even the stoic camel can suffer. If feeling safe, however, camels have been observed playing, rolling in sandy places and rubbing their legs and neck against other group members or trees, all signifying positive emotional wellbeing and a happy state of mind (Faraz et al., 2024). Nevertheless, play behaviour specific to camels has not been defined, in keeping with the lack of research on camels in general, especially their positive emotional states (Papageorgiou & Simitzis, 2022).

Information on stressful events such as during transport, overcrowding, mixing with unfamiliar camels or at slaughterhouses is more plentiful, especially related to debate around a camel's awareness of their impending fate as they smell blood and hear the distress calls of other camels (Kadim et al., 2024). Camels are most commonly slaughtered in the crouching position, but being forced into this position is unpleasant for the camel who may struggle intensely and make escape attempts, thus increasing the emotional contagion experienced by other camels awaiting the same fate.

Lynne McAlister, from Galahad's Animal Sanctuary in Victoria, Australia, knows firsthand the terror that spreads among these sentient animals when their freedom is taken away, and they are introduced to the world of humans via the horror of mustering, trapping, slaughter or enslavement. This story is reproduced with Lynne's permission.

Nestled amid nature's serenity, Galahad's Animal Sanctuary, run by Lynne McAlister, is home to rescued camels, horses, donkeys, cows, alpacas, sheep, pigs, chickens and dogs. Lynne's aim is to spread a message of hope and compassion, but also to inspire change through connection with the animals and hearing their stories. At Galahad's, each animal has a tale of resilience to share. They act as their own ambassadors, teaching people that simple acts of kindness and compassion, along with everyday choices, can have a significant impact on their lives. By 2025, Galahad's Sanctuary was home to 55 animals, although many more had been rescued over the years. Lynne confirms that her mission going forward is the rescue and provision of loving homes for more camels. As Lynne states, these majestic creatures often face neglect and hardship and need a voice which she has provided in the following segment.

Camels, my spirit animal

REPRODUCED WITH PERMISSION FROM LYNNE MCALISTER, GALAHAD'S ANIMAL SANCTUARY, AUSTRALIA

Camels are powerful spiritual symbols of adaptability, endurance and patience in many cultures. Their wisdom, ingenuity, intelligence and loyalty have been documented by all who live with them throughout history. They have adapted to survive desert-like environments and exhibit a high level of wisdom often overlooked by people who fail to see beyond their unusual physical characteristics and erroneous but widespread stereotype of aggression. In reality, they are wise beings, constantly observing their surroundings and learning various travel routes.

As an example, Lynne provided an excerpt from a book by Australian woman Robyn Davidson who travelled 2,700 kilometres from Alice Springs to Hamelin Pool in Western Australia in 1977. Travelling with her dog and four camels, Davidson had plenty of time to get to know her four unique and personable camel companions. Davidson's book *Tracks* (1980) documents this amazing journey and includes the following description of her camel travelling companions:

> They are the most intelligent creatures I know except dogs and I would give them an IQ rating roughly equivalent to eight-year-old children. They are affectionate, cheeky, playful, witty, yes witty, well-possessed, patient, hard-working, and endlessly interesting and charming. They are also very difficult to train, being of an essentially undomestic turn of mind as well as extremely bright and perceptive. This is why they have such a bad reputation. If handled badly, they can be quite dangerous and recalcitrant.
>
> *(p. 29)*

Lynne continued with her story: As an animal lover and advocate, I am distressed to learn about the plight of camels in Australia. The thought of these beautiful animals being slaughtered makes my heart ache. When I heard about an auction for two young camels, Milkshake and Catapult, I wanted to make sure they were safe from the "doggers". The doggers are notorious for buying horses and camels at auctions and selling them on to a knackery. They are then slaughtered and processed, often for pet food. I could not let that happen to Milkshake and Catapult. They were only two years old and had their whole lives ahead of them. I was determined to win the bid and give them a better future. At the auction, I learnt that they had originally been captured in the Northern Territory (Australia) and taken to a slaughterhouse. A Victorian property owner had them transported to his home to ride. Shortly after he fell ill and the property was sold at auction, including Milkshake and Catapult.

When I arrived at the auction, I saw Milkshake and Catapult in their pen. Milkshake was outgoing and greeted everyone who approached her with camel kisses touching her nose to theirs. Catapult, on the other hand, was much more reserved and quite anxious. I later learnt that he suffered from trauma due to being mishandled in his capture, making him very timid and fearful. But what struck me the most was how inseparable the two camels were. They were affectionate and seemed to draw strength and comfort from each other's presence. I knew then that I could not separate them—they had to stay together.

Winning the bid and getting them home was the easy part. I soon realised that Milkshake dealt with her fear of past abuse in a different way to gentle Catapult. Milkshake is a warrior queen and began asserting herself as herd leader, as many young camels would be in a new and scary situation. They are large animals and not without risks, so I approached a camel trainer for advice. As divine timing would have it, I found out they were about to hold a three-day camel training session. In this session, I helped to train a wild camel from the desert named Honey at a Camel Sanctuary and learnt so much about camels. Later, the same camel trainer came back to Galahad's Sanctuary and helped me train Milkshake and Catapult to halter, lead and hoosh.

Catapult had been so traumatised from his capture that he would stay far away. If you went near him, he could try and kick you in absolute fear. We spent a long time trying to stroke his face and desensitise him to touch so that he could learn to trust humans again. He was so stressed he would literally have diarrhoea, wail and vomit from us just stroking his face gently. Several years on, you can touch him and he loves kisses too. Milkshake has also had her own journey to feel more loved, secure and trusting.

But my journey with camels did not end there. A year later, I received a call from a camel dairy farm that was shutting down. They had a camel named Harley Quinn who needed sanctuary. They knew of my love for camels and reached out to me. I could not say no. After the camel dairy farm shut down, 240 camels went to auction. Harley was very small, did not produce much milk, and they were worried that she would not fare well at auction.

The more I learnt about camels, the more amazed I was by their intelligence and sensitivity. They are deeply misunderstood creatures, and I am determined to change that. I started educating people about camels and their plight. Harley Quinn, Milkshake and Catapult have become ambassadors for their species and show the world the true nature of camels. They are loving, gentle and intelligent animals who deserve our kindness and compassion. As I watch them play and roam freely during the day, I know I made the right decision. I had saved three lives and made a difference in their world, and that was all that mattered. I am now planning to rescue and rehome more camels.

Running an animal sanctuary and providing lifelong care to all residents requires ongoing financial resources for the expected—and unexpected—events that can arise daily. Fundraising goes hand-in-hand with animal care, whether for big items like a much-needed cattle crush, or veterinary care for animals still recovering physically and psychologically from trauma and abuse, or simply to feed a growing number of happy, healthy residents. Galahad's is no exception in their endeavours to keep making a significant difference to the lives and wellbeing of all residents.

ETHICAL IMPLICATIONS

David Mellor's Five Domains of animal welfare make it very clear: welfare-enhancing positive affects are fundamental to welfare. It is easy to identify welfare-compromising negative affects inflicted on camels, some of which are being addressed at a national or global level. However, so diverse is the camel's commodification by humans, there remain many areas that require urgent rectification, including tourism, entertainment, intensive farming, transport and slaughter. Welfare-enhancing experiences for these majestic stoic animals are harder to identify, and in fact, seem to be sorely lacking in many human and geographic environments. An obvious area for refinement for any roaming grazing animal is the provision of sufficient space to exercise their natural behaviour to walk, a requirement specified in Domain 4. This domain draws attention to

Figure 13.3 Harley Quinn, Milkshake and Catapult roam freely at Galahad's Animal Sanctuary during the day. (Image supplied by Galahad's Animal Sanctuary and reproduced with their permission.)

interactions with the environment (choice, free movement, exploration, foraging), interactions with other animals (bonding, rearing young, play, absence of threats) and positive interactions with humans (empathic, calm, clear).

Camels are truly miraculous animals, from their fat storing humps to their ability to chew even the thorniest bush and their third eyelid to protect from dust and sand, and yet humans have sought to harness these abilities for their own gain. As environmental changes move to favour the hardy camel, it becomes essential that their capacity to feel is not overlooked in the relentless drift towards commodification.

In 2025, an Australian Broadcasting Commission (ABC) television series, inspired by the Tasmanian Museum of Old and New Art (MONA) artist and curator Krisha Kaechele's book *Eat the Problem*, inferred that the problem of invasive pest species—including camels in Australia—could be resolved by encouraging more people to eat them. Titled *Eat the Invader* and prefaced with claims that the problem of invasive animals could be turned into an opportunity further demonises the feral camel and raises ethical implications around capture and slaughter of an animal whose commodification has already seen it used and abused globally. The title is also a timely reminder that the camel is not an "invader" of Australia—camels were brought into the country, discarded when superseded and ruthlessly killed when they impacted commercial livestock production and native wildlife. Human language is an important tool to allay the cognitive dissonance associated with the abusive treatment of some animals, and labelling camels as invaders justifies further mistreatment of these sentient beings.

REFERENCES

Abdullah, H.M. & Sheikh, A. (2024). History of a dairy camel (NAQA) and machine milking in Saudi Arabia; challenges and prospective. *Tropical Animal Health and Production, 56*(8), 352. https://doi.org/10.1007/s11250-024-04211-8. PMID: 39448450; PMCID: PMC11502539.

Ahearn, A. & Chatty, D. (2024, April 6). Turning camels into cows: mega-farms are being set up to produce camel milk on industrial scales. *The Conversation*. https://theconversation.com/turning-camels-into-cows-mega-farms-are-being-set-up-to-produce-camel-milk-on-industrial-scales-223485#:~:text=The%20camel%20may%20be%20the%20next%20cow.%20An,farming%20that%20produced%20modern%20cows%2C%20sheep%20and%20pigs.

Al Aiyan, A., Balan, R., Gebreigziabiher, S. et al. (2024). Comprehensive mapping of the exterior architecture of the dromedary camel brain. *Scientific Reports, 14*, 2971. https://doi.org/10.1038/s41598-024-53541-y

Al Jassim, R. (2024). Good feeding: Nutrition and feeding of the Arabian camel (*Camelus dromedarius*). In B. Padalino & B. Faye (Eds.), *Dromedary camel behavior and welfare*. Animal Welfare, *Volume 24* (pp. 55–70). Springer. https://doi.org/10.1007/978-3-031-48600-5_4

Alwahaibi, I., Dauletova, V., & Wels, H. (2023) Camels in the Bedouin Community of Oman: Beyond the human–animal binary. *Anthrozoös, 36*(1), 1–14. https://doi.org/10.1080/08927936.2022.2084993

Amandykova, M., Dossybayev, K., Mussayeva, A., Saitou, N., Zhunusbayeva, Z., & Bekmanov, B.A. (2023). Study of the genetic structure of hybrid camels in Kazakhstan. *Genes, 14*, 1373. https://doi.org/10.3390/ genes14071373

Benaissa, M.H. & Iglesias Pastrana, C. (2024). Good health: Recognition and prevention of disease and pain in dromedary camels. In B. Padalino & B. Faye (Eds.), *Dromedary camel behavior and welfare*. Animal Welfare, *Volume 24* (pp. 111–143). Springer. https://doi.org/10.1007/978-3-031-48600-5_6

Botox injected into camels to participate in Saudi beauty contest. (2021, Dec 08). *CE Noticias Financieras*. Retrieved from https://www.proquest.com/wire-feeds/botox-injected-into-camels-participate-saudi/docview/2608385785/se-2

Choudhary, M. (2021). Changing area, shrinking spaces and struggling species: *History of camels. History and Sociology of South Asia, 13*(2), 93–111. https://doi.org/10.1177/22308075211043285 (Original work published 2019)

Davidson, R. (1980). *Tracks*. Jonathan Cape.

El-Dwairi, Q.A., Al-Hussain, S.M., Banihani, A.S., Bataineh, Z.M., Djouhri, L., Mustafa, A.G., & Zaqout, S. (2023). Neuronal cell types in the spinal Trigeminal Nucleus of the camel brain. *Brain Sciences, 13*, 312. https://doi.org/10.3390/ brainsci13020312

Fallon, E., Krebs, G., & Bhanugopan, M. (2020). Survey of Australian dairy camel farms. *Journal of Camelid Science, 13*, 22–39. https://www.isocard.net/en/journal

Faraz, A., Hussain, S.M., Iglesias Pastrana, C., & Zappaterra, M. (2024). Good housing: Camels and their interaction with the environment. In B. Padalino & B. Faye (Eds.), *Dromedary camel behavior and welfare. Animal Welfare, Volume 24* (pp. 71–109). Springer. https://doi.org/10.1007/978-3-031-48600-5_5

Farm Transparency Project (2020). *Camels.* https://www.farmtransparency.org/kb/food/camels

Kadim, I.T., Al-Amri, I.S., AlKindi, A., & Al-Ajmi, D.S. (2024). The welfare of dromedary camels at slaughter. In B. Padalino & B. Faye (Eds.), *Dromedary camel behavior and welfare. Animal Welfare, Volume 24* (pp. 231–250). Springer. https://doi.org/10.1007/978-3-031-48600-5_11

Masebo, N.T., Zappaterra, M., Felici, M., Benedetti, B. & Padalino, B. (2023) Dromedary camel's welfare: literature from 1980 to 2023 with a text mining and topic analysis approach. *Frontiers in Veterinary Science, 10*, 1277512. https://doi.org/10.3389/fvets.2023.1277512

Menchetti, L., Faye, B., & Padalino, B. (2021). New animal-based measures to assess welfare in dromedary camels. *Tropical Animal Health and Production, 53*(6), 533. https://doi.org/10.1007/s11250-021-02978-8. PMID: 34739606; PMCID: PMC8568688.

Monsó, S. & Hintze, S. (2023). For their own good? The unseen harms of disenhancing farmed animals. In C. E. Abbate & C. Bobier (Eds.), *New omnivorism and strict veganism: Critical perspectives* (pp. 201–208). Routledge.

Narayanan, Y. (2021). 'A pilgrimage of camels': Dairy capitalism, nomadic pastoralism, and subnational Hindutva statism in Rajasthan. *Environment and Planning E, 6*(2), 756–775. https://doi.org/10.1177/25148486211062005

Nature (2020). *Camel fact sheet.* https://www.pbs.org/wnet/nature/blog/camel-fact-sheet/

Of botox and camel beauty contests: How a Saudi cultural festival generated global headlines. (2018, Jan 26). *TCA regional news.* https://www.arabnews.com/node/1233261/%7B%7B; https://www.arabnews.com/node/1233261/%7B%7B

Padalino, B., Aubé, L., Fatnassi, M., Monaco, D., Khorchani, T., Hammadi, M., & Lacalandra, G.M. (2014). Could Dromedary camels develop stereotypy? The first description of stereotypical behaviour in housed male Dromedary camels and how it is affected by different management systems. *PLoS ONE, 9*(2), e89093. https://doi.org/10.1371/journal.pone.0089093

Padalino, B. & Menchetti, L. (2024). The first protocol for assessing the welfare of dromedary camels (*Camelus dromedarius*) kept under nomadic pastoralism. *Frontiers in Veterinary Science, 11*, 1416714. https://doi.org/10.3389/fvets.2024.1416714

Papageorgiou, M. & Simitzis, P.E. (2022). Positive welfare indicators in dairy animals. *Dairy, 3*, 814–841. https://doi.org/10.3390/dairy3040056

PETA (2025). *What happens before and after camel rides*? https://investigations.peta.org/camel-rides-in-egypt/

Previti, A., Guercio, B., & Passantino, A. (2016). Protection of farmed camels (*Camelus Dromedarius*): Welfare problems and legislative perspective. *Animal Science Journal, 87*(2), 183–189. https://doi.org/10.1111/asj.12446

Ramadan, S., Nowier, A.M., Hori, Y., & Inoue Murayama, M. (2018). The association between glutamine repeats in the androgen receptor gene and personality traits in dromedary camel (*Camelus dromedarius*). *PLoS One 13*(2), e0191119. https://doi.org/10.1371/journal.pone.0191119

RSPCA (2024). What is the RSPCA's view on camel tour rides? *RSPCA Knowledgebase*. https://kb.rspca.org.au/knowledge-base/what-is-the-rspcas-view-on-camel-tour-rides/

Spencer, P.B.S., Hampton, J.O., Pacioni, C., Kennedy, M.S., Saalfeld, K., Rose, K. & Woolnough, A.P. (2015). Genetic relationships within social groups influence the application of the Judas technique: A case study with wild dromedary camels. *The Journal of Wildlife Management, 79*(1), 102–111. https://doi.org/10.1002/jwmg.807

Volpato, G., Dioli, M., & Di Nardo, A. (2017). Piebald camels. *Pastoralism, 7*(1), 3. https://doi.org/10.1186/s13570-017-0075-3

14

Donkeys and mules: Sentient animals, not beasts of burden

INTRODUCTION

"End of the road. Nothing to do and no hope of things getting better", muttered, Eeyore, the famous sad donkey from A.A. Milne's *Winnie the Pooh* series in the 1997 Disney film and book *Pooh's Grand Adventure: The Search for Christopher Robin*. For many donkeys, these words echoed the reality of their lives in servitude to humans. Their stoic steadfastness saw them commodified as beasts of burden, sold in meat markets when too old and sick to work, farmed for their skins to make traditional medicines, exploited as beach donkeys or in petting zoos before being abandoned or confined in intensive breeding systems.

But Eeyore also had moments of gratitude, as summed up in another of his favourite phrases "Thanks for noticin' me", and donkeys have increasingly been noticed and their welfare tackled at a global level. Grassroots welfare organisations sprang up to address the welfare of these sentient animals whose needs had been overlooked for centuries. The Flicka Foundation Donkey Sanctuary in the United Kingdom is one such organisation offering retirement and a safe home for life to donkeys rescued from abuse, neglect and abandonment. Other farmed animal sanctuaries such as Magzalea Farm Sanctuary in the USA include donkeys and a mule in their inter-species group of rescued and rehabilitated animals.

The global populations of donkeys (around 50.5 million) and mules (around 8.5 million) rival that of horses at 58.8 million, and yet they continue to receive less public and research attention to their welfare compared to their equid relatives (Bukhari et al., 2024). Researchers Syed Bukhari and colleagues question whether this relates to a "global north bias" and the tendency in animal welfare studies to concentrate on animal groups predominantly in the Global North. This includes cattle, sheep, pigs and horses as opposed to working animals such as donkeys and mules who are clustered in greater numbers in the Global South, especially Ethiopia, Sudan, Pakistan, Chad and Mexico (World Population Review, 2025).

DOI: 10.1201/9781003518853-15

In low- and middle-income countries, donkeys and mules play an important role in the lives of an estimated 600 million people worldwide as they carry loads such as bricks many times their weight, provide power to plough fields or pull carts where no mechanical means exists and provide transport in poor and marginalised communities (Bukhari et al., 2024). In the southern regions of Europe, donkey and mule trains are common, especially for tourism when they are called on to transport thousands of tourists and their luggage from visiting cruise liners up very steep cliffs to the towns (Thiemann & Foxcroft, 2016). Industrialised countries have largely replaced donkeys with mechanised engines, but they remain indispensable in many poorer regions, especially in hot and arid climates and in places where transport infrastructure is lacking.

The global donkey population has been steadily increasing since the early 1960s, and yet the economic importance and the welfare needs of these hard-working animals are often underestimated (Maśko et al., 2022). In contrast, this increase in donkey numbers has been countered by a decrease in the number of mules between 1997 and 2018 (Norris et al., 2021). Interest in donkeys for non-working roles such as meat and milk production, cosmetics, nutraceuticals and onotherapy (the donkey equivalent of equine therapy) has turned the focus to both their welfare and the cruelties inflicted on them in some situations. Overwork, neglect, poor traditional husbandry practices and methods of slaughter have taken a toll on donkey welfare globally.

SOME BACKGROUND HISTORY

The initial domestication of donkeys (*Equus asinus*) reportedly occurred more than 7,000 years ago, approximately when the Sahara first became a desert (Gross, 2022). Evidence suggests that domesticated donkeys descended from two sub-species of the African wild ass (*Equus africanus*)—the Nubian wild ass (*Equus africanus africanus*) and a population in modern Eritrea (*Equus africanus dianae*), both of which are now presumed extinct in the wild (Murray et al., 2013).

Bones and teeth from Sumerian sites (now south-central Iraq) were linked to the African wild ass morphology rather than the other possible ancestor, the onager (*Equus hemionus*) which, when bred with African wild asses, results in sterile offspring. However, the assumption that donkeys originated solely from this region with the domestication of wild asses has been questioned with remains from early sites in western Asia identified as the African wild ass. This led to suggestions that there were two centres of domestication of the donkey, both in northeast Africa (Clutton-Brock, 2012). Subsequent analyses of donkey genomes since the start of recorded donkey history found evidence for two competing hypotheses; that is, a unique origin in Africa and dual origin in separate locations in Africa and the Arabian Peninsula (Gross, 2022). However, the eventual conclusion supported the unique African origin as the most viable based on modern donkey genomes, with an even spread into the Arabian Peninsula and Eurasia.

Donkeys have been working for humans a lot longer than horses, pulling chariots in ancient Mesopotamia and even providing a ride for Jesus in the Christian tradition. They feature in ancient hieroglyphics, in Aesop's tales, the Quran and other cultural artefacts (Gross, 2022). They also played a ritualised sacrificial role, with their skeletons found in burial grounds such as the Akkadian site of Tell Brak in Syria (2200 BCE) (Clutton-Brock, 2012).

Equus is a genus of mammals within the family *Equidae* that includes domestic horses (*Equus caballus*), donkeys and horse/donkey hybrids known as mules (Norris et al., 2021). Mules arose once the spread of horses caught up with donkeys around 4,000 years ago, and humans started deliberately creating hybrids to provide better work animals. There was a perception that this first-generation hybrid from a male donkey and female horse, although infertile, was a more patient and useful work animal than both its parents (Gross, 2022). When a female donkey bred with a male horse, a hinny was born.

Donkeys are quite distinct from horses, with variations in anatomy (digestive tract, spine, skull and peripheral skeleton), social behaviours, physiology (ears, neck and feet), dietary requirements and susceptibility to disease (Maśko et al., 2022). Their territorial nature sees them stand their ground with other species, compared to horses who are dependent on the flight defence response. Donkeys also have lower nutritional requirements compared to horses, allowing them to forage on lower quality grounds but also putting them at risk of obesity and metabolic disease if housed and fed like horses.

DONKEYS AND MULES IN CONTEMPORARY SOCIETY

Fast forward to the twenty-first century, and globally, the stoic donkey plays a variety of roles in some of the poorest communities, often to the detriment of their wellbeing. Although indispensable to the livelihood of some people, they experience daily welfare risks when forced to transport loads many times their weight. Carrying anything from waste removal to bricks to heavy tourists and their luggage, and compounded by poor handling and whipping, lameness, dehydration, heat stress, exhaustion, malnutrition and injuries, a donkey's life can be short and miserable. The erroneous belief that widening a donkey's nostril would relieve their respiratory distress during heavy work in hot climates added an additional level of pain and suffering when owners cut a vertical slit from the ventral aspect of the nares up into their donkey's nostril (Shah et al., 2019). A study in India reported nose-splitting, ear-splitting and branding to be common wounds among donkeys, although misperceptions about nose-splitting were being addressed through community interventions and the support of key community leaders. Festering sores from poorly fitting harnesses rubbing in the harsh environmental conditions were common and compounded by a lack of supporting infrastructure such as veterinary care, welfare legislation and regulatory bodies (Bukhari et al., 2024). When no longer capable of working, the demand for donkey skins to produce the traditional medicine "ejiao" guaranteed a miserable ending for these patient sentient animals (Norris et al., 2021).

While donkeys may not experience the factory farm conditions of other domestic animals such as cows, pigs and chickens, the range of cruelties inflicted on them is diverse and shocking. Whether pulling carts of bricks in Egypt or human waste in Pakistan and Ethiopia, several studies have consistently shown the prevalence of lameness and wounds among working donkeys to range from 42 per cent to 96 per cent in some regions (Farhat et al., 2020). The differences largely related to environmental conditions, type of work and harness system.

In Egypt, where over three per cent of the global donkeys and mules have been toiling for humans for thousands of years, human population growth and increased building construction to provide adequate housing have negatively impacted these stoic animals (Farhat et al., 2020). Brick kiln donkeys who cart bricks between kilns and building sites have the highest rate as well as the most severe lesions from ill-fitting unpadded harnesses. Poorly designed or constructed harnesses create an inefficient transfer of power from the donkey to the cart, intensifying fatigue, discomfort and injury. When adding poor nutrition and hydration, the resultant physical misery of these donkeys and mules is evident in apathy, depression and, at times, aggression.

In Pakistan, which has the third largest donkey population in the world (Shah et al., 2019), donkey carts play an essential role in the transportation of building materials, fresh produce and human garbage. One study identified that donkeys in Karachi, Pakistan, are indispensable in both management of the 14,000 tonnes (15.4 tons) of daily solid waste produced by the 20 million residents and providing a livelihood for donkey owners and their families. Donkey cart owners reported that their donkeys transported on average more than 1,000 kilograms (2,204 pounds) of non-recyclable and recyclable waste daily. With male donkeys weighing 500–600 kilograms (1,102–1,322 pounds) and females 450–550 kilograms (992–1,212 pounds), the toll on their physical and mental health was relentless.

Ejiao demand

Those donkeys not destined to pull heavily laden carts until they dropped could instead find their hides in high demand for Chinese medicine. For centuries, ejiao, a glutinous material extracted from donkey hides, has been valued as a means to improve strength and health, deliver youth-preserving properties and assist some psychological issues including insomnia (Bukhari et al., 2024). The slaughter of donkeys in China to meet the demand for this luxury item significantly reduced their numbers, prompting an increase in donkey farming in China and donkey slaughterhouses in countries such as Africa and South America to maintain supply. Large-scale trading—both legal and illegal—and resultant handling, transport and slaughter in an unregulated environment had devastating implications for these donkeys' welfare (Lima et al., 2022). Donkeys faced potential mistreatment at every step, condemned to spend many hours confined during long journeys with limited or no access to food and water and subject to non-compliant slaughter practices when they arrived. Almost two-thirds of the world's donkeys live in Africa, and high demand and prices for skins saw some local African communities, dependent on working donkeys for their livelihood, lose animals through theft. Families were left destitute, and donkeys experienced relentless cruelty in the name of human greed. Other reports suggested that the export of feral donkeys from Australia also provided a source, as the combination of high demand and shortage made ejiao a very lucrative market (Goodrum et al., 2022).

In February 2024, People for the Ethical Treatment of Animals (PETA) provided an update on their ongoing campaign to prevent the slaughter of donkeys in Africa: "Following multiple investigations and years of campaigning by PETA entities […] African Union leaders have approved a ban that makes the slaughter of donkeys for their skin *illegal* across the entire continent for the next 15 years. This moratorium on donkey slaughter deals a massive blow to the Chinese *ejiao* industry […] and will prevent millions of donkeys from enduring painful, terrifying deaths" (PETA, 2024).

Milk and meat

Donkeys who were not carting bricks or human waste until they dropped or being skinned for the valuable gelatinous substance in their skin risked being commodified in another way. To support the growing demand for donkey milk in parts of the world, including Europe, for babies allergic to other types of milk and for cosmetic production, industrialised production emerged. This ranged from intensive to semi-intensive and semi-extensive farming systems (Dai et al., 2019). With limited specific legislation or best practice guidelines to protect donkeys and considerable variability on different farms, some donkeys were subjected to poor welfare outcomes, especially from inexperienced producers (Thiemann & Foxcroft, 2016). Dairy donkeys, like cows, were obliged to produce extra milk for human consumption, therefore good feeding played an essential role in their welfare (Raspa et al., 2019). However, when donkeys were fed twice a day as was the norm in traditional intensive farm systems, problems could arise for this single stomach, hindgut fermenter animal who was more adapted to "trickle feeding" of small amounts continuously (Bukhari et al., 2024).

The unknowns in donkey care and welfare, including opportunities to express natural behaviours, posed ongoing risk factors as modes of production intensified to meet demand. Intensive farm settings deprived donkeys of the opportunity to form lasting attachments when they were moved around and separated, triggering potential physiological and psychological stress (Thiemann & Foxcroft, 2016). Concerns about consistent levels of care and welfare point to a need for industry standards and guidelines—including the suffering of donkeys leading up to and during the slaughter process—and serious consideration of whether intensive farming is appropriate for these sentient animals.

Tourism, companion and therapy donkeys

There is yet another cruelty that humans inflict on the long-suffering donkey. By the early twentieth century, donkeys were increasingly associated with and exploited for tourism around the world. Whether it was providing transport to visit ancient ruins or similar expeditions, lugging cruise liner passengers and their luggage from the dock up steep cliffs to picturesque towns, poked and prodded in unregulated children's farm attractions or providing children's rides at seaside resorts around the world, donkeys made a significant contribution to the global tourism industry despite limited research and regulation around their welfare needs (Tully & Carr, 2020).

In 2022, PETA Germany released video footage of the continued exploitation of donkeys for tourism on a Greek island. Despite some reforms following a similar expose in 2017 and demonstrations outside Greek embassies around the world, PETA claimed exhausted dehydrated donkeys and mules with bloody wounds continued to carry heavy loads of tourists up and down the steep slippery steps from cruise ships to accommodation or sightseeing trips (PETA Australia, 2023). PETA's report included commentary from equine expert Dr Maximilian Pick who reviewed the photos and video footage and concluded that the donkeys and mules exhibited numerous, partly untreated skin wounds and scar tissue, and affirmed the cruelty of having animals stand in the blazing sun without water or food. PETA, along with groups such as the Donkey Sanctuary and the Society for the Protection of Animals Abroad (SPANA), remain determined to raise awareness of the poor treatment of donkeys across the tourism industry globally (Tully & Carr, 2020).

Donkeys kept as companion or therapy animals experience a completely different lifestyle to the exhausted tourism donkeys, but there can still be challenges. With an average lifespan of around 40 years, a donkey can sometimes outlive the caring capabilities, or the longevity, of their human and find themselves alone and abandoned. Similarly, the growing movement towards using donkeys as therapy animals offers physical benefits compared to their overworked relatives, but there can be emotional welfare implications if failing to recognise that every donkey is unique, and a calm behavioural response may belie underlying stress (Clancy et al., 2022). Their stoic nature and limited visual cues as to a donkey's psychological wellbeing can magnify emotional risks, while serious health issues such as obesity, diabetes or hoof problems when housed and fed with horses posed ongoing physical risks. Donkeys adapted to arid areas with limited vegetation, so when tempted with the lush abundance of more temperate regions, colitis or laminitis (inflammation of the foot) could result. The donkey's minimisation of overt reactiveness meant these painful conditions could go undetected and untreated (Maśko et al., 2022).

Even donkeys who escape, or are deliberately released or abandoned, experience welfare issues when their environmental resilience facilitates the establishment of feral populations. In Australia, where feral donkeys achieved official pest status in 1949, some are mustered or trapped and sold for pet meat, but the majority face aerial shooting from helicopters in an attempt to control their numbers (NSW Department of Primary Industries, 2022). As with feral camels in Australia, there is a failure to acknowledge that donkeys were brought into the country, discarded, then ruthlessly hunted and killed when they became a commercial and environmental liability.

FEELINGS AND EMOTIONS

As sentient creatures, there is no doubt that donkeys feel all the negative experiences inflicted on them, but their complex cognitive abilities and emotional capacity may not be evident when subverted by human control. When PETA released a poignant image of a Greek tourist donkey, with his eyes closed as he leant against rocks, seeking brief respite from the hot sun and the next load of heavy tourists, the donkey's posture and facial expression clearly depicted apathy, depression and exhaustion (PETA Australia, 2023). However, he would soon be forced back to work in a clear act of speciesism inflicted by humans oblivious to the suffering in his body language. For decades, psychologists have asserted that human body language could convey more than words, knowledge that is crucial in the therapeutic process. While the donkey's stoic nature reduced overt visual clues as to their emotional experience, the body language of this Greek tourist donkey was impossible to miss—to anyone who cared to look. Quoting A.A. Milne's Eeyore again, this defeated donkey's attitude could be easily summed up with the words: "I never get my hopes up, so I never get let down" (The Book of Pooh, 2001).

Like all mammals, donkeys and mules have neural substrates that leave no doubt as to their ability to feel pain and suffering as well as positive emotions, if given the opportunity. Their neocortex with extensive gyration (folding) supports intelligence and emotion, while their limbic system (ancient part of the brain) provides the relevant structures associated with emotional capacity. In other words, they are sentient beings capable of experiencing positive and negative emotions in response to stimuli in their life (Lima et al., 2022). Olfaction (sense of smell) plays a key role in donkey social interactions, especially between mothers (jennies) and their new-born foals (Merkies et al., 2020). Mother/infant bonding elicits changes in neural structures affecting the olfactory bulb, driving maternal recognition and the jenny's memory, both crucial to her baby's survival. The amygdala, part of the limbic system and therefore instrumental in emotional capacity, processes the olfactory stimuli of the newborn, and memory performance is continually reinforced through odours. Emotions elicited through aversive olfactory stimuli are similarly processed in the amygdala, making this brain structure important for combining memory and emotion in associative learning, or remembering and learning through experience. Another part of the donkey's brain associated with memory—and therefore emotion—is the hippocampus. The hippocampus also responds to olfactory stimuli and has a role in long-term memory, processing emotional experiences and responses to stress. The odours associated with the many roles forced upon donkeys would see their brain structure adjusting accordingly to increase heightened emotional responses.

Donkeys have sensory receptors that respond to painful sensations. The ill-fitting harnesses when pulling bricks, or the festering lesions from slipping on garbage when hauling heavy loads, or the unregulated methods of slaughter by bludgeoning or stabbing, all result in a nociceptive response that is transmitted to the brain and interpreted as pain. While pain may be inferred from the physical wounds on a donkey, among companion and pleasure donkeys, it may be less evident, given their typical stoic nature and temperament.

Pain is a physical stressor, but the events in a donkey or mule's life that create pain can also induce fear, which is a psychological stressor. Fear in a natural setting facilitates survival by triggering metabolic changes associated with the "fight or flight" response. This prepares an animal to take action or retreat, after which the stress response subsides. However, when exposure to a fearful event occurs outside of an animal's natural environment and is long-term or inescapable, this can set in motion chronic metabolic changes that are detrimental to wellbeing and survival. Among donkeys and mules, relentless beatings to submit or continue working past the point of exhaustion, or confinement with unfamiliar animals during transport, auction or slaughter, can all trigger chronic fear that is no longer a valid survival mechanism (Lima et al., 2022).

Maeve the mule experienced physical pain—that was evident from the marks on her body. Maeve's behaviours suggested she had also experienced fear before being rescued from the kill pen at an auction, but that would never be known for sure. Removing the source of physical pain from Maeve's body was only the start of her journey to a life without fear within the safety of Magzalea Farm Sanctuary in New Hampshire, USA.

Maeve the mule's life without fear and pain—Magzalea Farm Sanctuary, USA

Halters are for safe handling and leading of an animal, but they need to be suitable for the animal's size, behaviour and purpose. It was clear that Maeve's halter, so tight it was embedded in her skin and needed cutting off, had not met these requirements for many years.

Maeve's face and expression saved her life. The ominous words beneath her captivating image— "Shipping to slaughter February 5, 2019"—could not be ignored, even when the sanctuary was temporarily not accepting new residents. Everyone in rescue knows there is a "line" where no more animals can be accepted until funding reaches a certain point, or long-term projects are fulfilled. And everyone in rescue knows there is always that one animal who slips across the line. Maeve was that animal, but unbeknownst at the time, she would also bring along a pregnant friend Alice.

Maeve was bailed from the kill pen and awaiting transport to Magzalea when a concerned woman contacted the sanctuary to say that she had paid the bail of a six-year-old Shetland pony who now had nowhere to go. Maeve the ex-working mule and Alice the possibly pregnant Shetland pony crossed the line together and began their very different journeys to rehabilitation and security.

For Maeve, separating the halter growing into her head and healing the signs of past physical and psychological mistreatment would prove challenging when she could not be touched, nor even looked at on some days. Whether she could ever learn to trust a human after being beaten and worked until her spirit was broken, then discarded like a piece of equipment that was no longer useful seemed unlikely at times, but giving up was not an option.

Mules are sentient animals with capacity for both positive and negative feelings about their experiences. When negative experiences outweigh positive experiences, a psychological phenomenon called cognitive bias can develop where emotion influences cognition and future ambiguous stimuli are more likely to be judged as negative (Nguyen et al., 2020). Cognitive bias has been reported among some animals, making it possible that for Maeve, vet visits, vaccinations, halters—even looking at a halter—were all viewed through the lens of her past negative experiences as she struggled to adjust to human compassion and care.

However, what a difference kindness, positive reinforcement and patience made to Maeve's view of her new world and the humans in it. Within a month, she had shown signs of learning to trust, to give and receive love and to feel safe

Figure 14.1 With time, patience and positive reinforcement, Maeve learned to trust again. (Image supplied by Magzalea Farm Sanctuary and reproduced with their permission.)

in her new forever home. Importantly, Maeve could finally display the unique personality that had been quashed in her earlier life. Fun-loving, affectionate and appreciative of the simple things in life like a good brush, Maeve progressed to shoving everyone out of the way during sanctuary tours to gain sole attention from visiting humans who she no longer feared or mistrusted—so different from the girl who could not bear to be looked at a few short years ago. Safety allowed Maeve to display both her sensitivity and strong will, but she never failed to respond to positive reinforcement reminding her of boundaries in the shared interspecies space of Magzalea Farm Sanctuary.

And what of Alice? Maeve and Alice developed a strong friendship on arriving at the sanctuary, but Alice's journey to feeling safe in her new home would take longer. Alice had never been handled or treated with care, instead viewed as a "baby maker" and left out in a field with minimal interaction. Alice was pregnant and gave birth to a son, Eddie. For the first time in her life, Alice was able to keep her foal, and five years later, their bond remained unbreakable. But there is always room for her best friend Maeve, along with Jethro, the mini mule who Alice adopted, and Patrick, a big mellow mule.

There is no doubt that in the minds of some people, the horse has higher status than the donkey or mule who struggle to shake off false stereotypes associated with stubbornness and stupidity. The donkey's tendency to resort to either fight or freeze behaviours while they work out what to do in the face of fearful stimuli is in stark contrast to the horses' more dramatic flight response (Merkies et al., 2020). It has also added to the obstinate, stupid and unfeeling labels applied without consideration of their intelligence and capacity to experience a wide range of emotions (González et al., 2019). In fact, calm and cautious endurance may mask extreme stress as the donkey's past experiences, training and current environment all contribute to their decision-making and behavioural response (Clancy et al., 2022).

If donkeys do experience welfare issues beyond their physical ailments, this raises concerns about accurately assessing the feelings of such a stoic animal. Much of the research has focused on addressing the prominent physical issues associated with overwork, poor fitting harnesses and general neglect. But many working donkeys live in a constant state of fear and depressed mood linked to harsh working conditions and humans with insufficient knowledge, ability or desire to properly care for their working donkey (Farhat et al., 2020). Unfortunately, empirically measuring this stress requires more scientific research, the lack of which perpetuates some of the worst atrocities occurring in regions where researchers may be reluctant or not welcome to spend time. To improve donkey welfare requires action by those perpetuating the welfare breaches, but roadblocks develop where this action impinges on a family's income and a culture's beliefs.

Researchers Samantha Seganfreddo and colleagues (2023) assessed ear position as an indicator of positive and negative feelings in donkeys. They reported that donkeys' ears were held for a longer time in the upright frontal position during apparent negative emotional states, similar to other studies that reported ears in a frontal position to be associated with greater alertness. Ears held sideways were more indicative of emotionally positive contexts. Donkeys in the study also tended to distribute their body weight on all four legs during a pleasant situation and kept one of their forelimbs bent for longer in a frustrating emotional situation. This was associated with leg lifting movements indicative of increased restlessness and agitation. While information on body language is crucial to assessing welfare, unfortunately it may be insufficient to prompt remedial action among those people living in poorer regions and dependent on their donkey for the family's livelihood.

Another indicator of emotional capacity can be found in pair-bonding and friendships, with some research reporting that domestic donkeys prefer a small number of strong friendships, which are long-lasting and situation-specific (Murray et al., 2013). These findings were based on some donkeys' clear preference for being close to one other individual and reciprocation of that preference, but the researchers also commented that individual factors such as age, temperament and history prior to arrival at the sanctuary, as well as the social dynamics of group composition and stability, could impact the formation of friendship bonds. While research continues to validate the importance of friendship bonds between donkeys, meeting some of the rescued donkeys at The Flicka Foundation Donkey Sanctuary in the United Kingdom quickly verifies that friendships can be fundamental to a donkey's emotional wellbeing.

The power of friendship—Matilda, Violet and Benji, The Flicka Foundation Donkey Sanctuary, United Kingdom

Located in the tranquil environment of Penryn in Cornwall, United Kingdom, The Flicka Foundation Donkey Sanctuary provides a stark contrast to the abuse, neglect or abandonment previously experienced by these sentient animals. Some were sick or elderly and of no further use to their human owners, others on their way to meat markets. Others had spent years in intensive breeding programmes where their foals were taken from them year after year until they were too old and ill to breed.

Since 1995, donkeys, horses and ponies have been cared for by Flicka's experienced team as they learn to enjoy environmental enrichment and, for the first time in their lives, freedom and choice (The Flicka Foundation Donkey Sanctuary, n.d.). Deciding how to spend their days can include grazing in a choice of paddocks, relaxing, grooming, playing, socialising with their friends or spending time in one of the large, cosy indoor donkey barns during colder weather. A growing number of the donkey residents were saved from certain death on the European continent, and gradually, the Flicka Foundation evolved into a predominantly donkey sanctuary in recent years.

Currently caring for over 100 donkeys, horses and ponies at the sanctuary, all are unique individuals with a story to tell and a future to enjoy. Importantly, they have the right to form friendships that remain stable for the rest of their natural lives, as Matilda and Violet's bond reveals. Their story, as well as Benji's story, has been provided by The Flicka Foundation Donkey Sanctuary and reproduced with their permission.

Matilda and Violet

Matilda arrived in the United Kingdom in October 2010, via Ireland, where she was being moved through the live animal markets destined for the continent and, very likely, the meat trade. She had a foal at foot (who she arrived with) and was very scared and unapproachable, showing defensive aggression and bearing physical scars from what we can only assume were beatings. Her ears were flattened back as we watched her herded off a lorry with a stick. Ear posture and body language can be indicators of emotional state in donkeys (Seganfreddo et al., 2023), and everything about Matilda revealed distress and trauma. When she arrived with us, her skin, feet and teeth were in a bad way, but she was terrified and her fear made it difficult to treat her. After initial assessment and carrying out what treatment we could to keep her comfortable, we housed Matilda with some other donkeys, including one called Violet.

Violet had arrived at Flicka with her Mum, Bluebelle, who was emaciated and in poor condition. She was just a young foal at the time and poor Bluebelle was actually carrying another foal. Violet was nervous of everything, having had a very traumatic start to her life. She had endured weeks in the back of a live transport lorry, destined for the meat market and the continent.

Sweet-natured Violet seemed to sense Matilda's distress and immediately went to her (whether Matilda liked it or not!). She stood with her and never left her side and has not left her side since, all these years later. Violet's soft, friendly nature really helped Matilda, who settled and began to trust us in just a few short weeks, turning into a loving and gentle girl.

The two girls are still utterly bonded and always together. We often talk about the physical state of donkeys when they arrive, but the emotional trauma they suffer goes deeper than we can possibly know and often stays with them for a lifetime. Such is their sensitivity and depth of emotions.

Benji

Benji arrived in spring 2023, just a baby, after we received a call to say he was living alone on a remote "second hand dealer" site. He had no paddock, no shelter and, when our caller saw him, he was trying to sleep huddled up on a patch of concrete in the pouring rain. He arrived with us soaked in urine and covered in his own droppings. Benji received a thorough health check, as well as treatment for rain scald and worms. He also had his feet trimmed and enjoyed some good, solid food, lots of hay and a deep straw bed.

Despite his terrible start in life, Benji is very friendly, cheeky and naughty! He is obsessed with wheelbarrows and loves to playfully stop us moving them, so it is hard to get far with our dung-picking when he is around! It is hard to comprehend the awful situation he was in, but now he is living the life a donkey should, with company, care and love.

Benji may join one of the larger herds of donkeys at our sanctuary in time, but for now he is very happy with his friends who arrived in the same year—Little Bear, Freddy, Robbie and Ned. The three youngest boys love to play and chase

Figure 14.2 Matilda and Violet remain bonded and always together. (Image supplied by Flicka Foundation Donkey Sanctuary and reproduced with their permission.)

Figure 14.3 Benji is friendly, cheeky and a bit naughty, but he looks so innocent here. (Image supplied by Flicka Foundation Donkey Sanctuary and reproduced with their permission.)

and are often seen running around in a line in size order all biting each other's bottoms—Freddy in the lead, Benji in the middle and Little Bear (quite literally!) bringing up the rear!

Friendship bonds under domestic conditions provide no obvious physical survival enhancing functions, suggesting instead they offer essential psychological and emotional benefits. Many donkeys are forced to live alone as working animals in lower-income countries, where the physical burden of overwork, pain and exhaustion can be compounded by the emotional toll of loneliness. With a potential longevity of 40 years or more, domestic donkeys need to live in groups of two or more for their overall wellbeing. Extreme distress, pining and loss of appetite can arise when these bonds are broken (Murray et al., 2013). Even short-term separations can result in anxiety, with mental and physical signs of distress including vocalising, pacing and flat character. Fortunately, it was one donkey's need for companionship that led to the rescue of two donkeys by Magzalea Farm Sanctuary.

Marigold and Petunia, mother and daughter—Magzalea Farm Sanctuary, USA

Donkeys need friends, and children need their mothers, two simple facts that led to mother and daughter donkeys Marigold and Petunia finding their forever home at Magzalea Farm Sanctuary.

When long-time donkey resident Poppy unexpectedly lost her best friend Dixie, she was devastated. Donkeys form strong friendships and can pine, vocalise and lose their appetite when bonds are broken. Learning that a donkey named Marigold needed a home seemed the perfect solution for both ladies, and so the process of getting Marigold to Magzalea Farm Sanctuary began. However, when discovering that Marigold had a timid six-year-old daughter named Petunia, the decision was easy—Petunia had to stay with her mother, meaning two donkeys instead of one would join Poppy.

Marigold and Petunia's earlier life events were unknown, but the "broken crests" on both ladies—formed by storing excess calories in fat deposits around their necks—suggested that they had been overweight at some point. After evolving in arid environments, donkeys suffer physically if fed the rich lush food typically given to horses rather than the fibrous plant material their digestive systems tolerate and require. Poor foot health is another indication of having been overweight and despite Petunia's extreme timidness, calling in the sanctuary's friend and farrier to relieve the pain and discomfort in both ladies' distorted feet could not wait for Petunia to settle in. Mum Marigold relished the pedicure, providing a wonderful role model to Petunia who was still working on trusting humans.

The next move was into the living quarters of Hilda and Sadie, two senior mini donkeys who had passed away a few years earlier after ten years of safety and peace at Magzalea. Hilda and Sadie were as tough as they were loving, leaving an indelible mark on all those who knew and respected them. In the peace and serenity of these living quarters, Petunia continued her journey to psychologically heal and feel safe enough to trust humans again. There was elation when Petunia accepted a brief head rub and allowed the "crusties" to be wiped from her eyes. When running a farmed animal sanctuary, moments like these are precious rewards for the time and patience devoted to each resident's unique journey to wellness. Mum Marigold had already realised that these humans were okay and proved to be accepting and forgiving.

Long-time resident Poppy was more than ready to take on the rescued mother/daughter donkey duo, creating a beautiful "bouquet" of donkey ladies. Six months later, another donkey—coincidentally named Daisy—added to this group of floral ladies. Daisy's anxiety on arrival would rival even Petunia's, preventing anyone getting close enough to remove her halter. Watching her new forever family accepting the humans encouraged her to quickly take a treat by hand. But just like Petunia, earning her trust required time, love, patience—and many treats.

Pedicure number two six weeks later saw Marigold even more relaxed and enjoying the moment, although insisting on being quite the helicopter mother when it came time for her timid daughter to step forward.

As Marigold's confidence grew, so too did her cheekiness. Stealing her daughter's breakfast was a regular challenge requiring distracting techniques such as a relaxing massage on her back by a willing human. With Petunia on a supplement and Marigold on Cushing's disease medication, stealing her daughter's food and mixing these medications was not an option.

Six months after arriving, both ladies were feeling more at home, to the point that Marigold had become overly demanding for nose rubs, ear and bum scratches. Now adjusted to her medication for Cushing's disease (a common hormonal disorder that affects the pituitary gland in the brain), Marigold's trust and attachment to humans was complete. In fact, personal boundaries did not exist for this loving donkey who was no longer constrained by the unknown events in her past that had left her daughter so wary of humans.

Petunia was also mellowing, taking treats by hand and leaving the fingertips behind, although mornings proved a challenge. Petunia was not a morning person and needed at least an hour after breakfast before any interaction was possible. Living at the sanctuary meant Petunia's morning grumpiness was accepted and respected as part of her unique personality, as were her love of nose rubs and the occasional neck rubs. Leg and feet touches were improving but still very mood dependent (and never in the morning).

One year on, little Petunia would often peek over her mother's back to ask for nose scratches. Petunia has learned to request whatever she wants and given the long journey she has been on to heal physically and emotionally, many would rush to accommodate her wishes.

ETHICAL IMPLICATIONS

Human culture and donkey and mule welfare have a long and interconnected history, with donkeys and mules occupying a range of complex spaces in all parts of the world. Animal rights charities from developed nations advocate for the

Figure 14.4 Mother and daughter, Marigold and Petunia, feel safe to express their unique personalities. (Image supplied by Magzalea Farm Sanctuary and reproduced with their permission.)

welfare of donkeys and mules in poorer countries where they can be overworked, underfed or skinned for their hide. However, donkeys and mules also face exploitation in developed nations where they may experience unrealistic demands in the tourism industry or neglect and abandonment.

Discussion around redesigning the harnesses for more comfort in load bearing donkeys misses the over-riding ethical and welfare issue: expecting donkeys to cart loads many times their weight is inhumane and unsustainable (Farhat et al., 2020). Interventions for donkey welfare should not rely solely on redesigned harnesses and providing veterinary treatment when the root cause remains the power imbalance in the human–donkey relationship. Donkeys are fundamental to some human livelihoods, increasing the urgency to develop a comprehensive cultural understanding of how this ongoing mistreatment can be mitigated.

Farmed animal sanctuaries support the individual donkeys and mules, but a large-scale global movement is essential to understand and remedy all facets of the donkey's life. In 1923, SPANA, a society for the protection of animals abroad, was born out of a United Kingdom mother and daughter's desire to relieve the suffering and neglect of many of the working animals they encountered during their travels across North Africa. Seeing malnourished donkeys, mules and camels collapsing under the weight of huge loads or with bloody painful wounds from poorly fitted harnesses compelled Kate and Nina Hosali to do something about this misery. The principles of "treating, training and teaching" underpin this United Kingdom-based international charity, which recognises the inextricable links between the welfare of working animals in low-income countries and the subsistence of humans in these areas.

Similarly, the Donkey Sanctuary, a United Kingdom-based equine charity, provides sanctuary, advocacy, education, research and on-the-ground practical support for millions of donkeys and mules globally (Thiemann & Foxcroft, 2016). With several European bases and associations with other equine/animal welfare charities, donkeys and mules benefit from veterinary care, the training of local vets and forged links with local community leaders, owners and paraprofessionals such as farriers, equine dentists and harness makers.

Iraklis (Hercules) and new beginnings—Agia Marina Donkey Rescue, Greece

With increasing public scrutiny and advocacy around the global plight of donkeys, a young stallion donkey named Iraklis (Hercules in English) exemplifies the dedication and teamwork required to save a single abused donkey. Identifying donkeys in need is only the starting point—they then require medical attention, routine care of teeth and feet, and somewhere to live out the rest of their lives in safety and security. Agia Marina Donkey Rescue in Crete, Greece is one such place offering sanctuary to rescued donkeys (Agia Marina Donkey Rescue, 2020).

A lone donkey's plight when a farmer could no longer care for her provided the catalyst for the establishment of Agia Marina Donkey Rescue in 2004. With some help from Caring for the Animals Trust [CARAT] UK, Animal Action Greece, and an experienced veterinarian, the family-run rescue was eventually able to offer sanctuary, food and medical care to the many humble, intelligent and unique donkeys fortunate enough to call Agia Marina their home.

Iraklis lived and worked on the tiny island of Iraklia, next to Naxos. In 2020, Iraklis collapsed under the weight of the load he was expected to carry every day, resulting in a severe beating from his owner. Ironically, this cruel beating provided the catalyst for Iraklis to escape his life of exploitation. The owner had beaten Iraklis in a public place, and when a kind citizen reported this to the authorities, the owner was prosecuted and Iraklis permanently removed from his care.

Fanis and Barbara from Agia Marina Donkey Rescue agreed to take him in, and with financial and travel assistance from CARAT, Iraklis journeyed overnight by boat to begin his new life in Crete. The lesions on Iraklis' hindquarters and neck from the heavy loads of bricks and stones he had carried his whole life and the regular beatings when he faltered, gradually healed and faded. But the psychological scars ran deeper as his fear of any human contact initially left him anxious and mistrustful.

Tasty food helps solve many problems, and it was clear from Iraklis' gusty enthusiasm that he had never experienced the sheer joy of carrots and sliced apples. He loved them! Teeth and feet were next, and despite this being a new experience for the neglected donkey, he seemed to know instinctively that the farrier and dental technician were there to care for him as he stood patiently while his sore feet were trimmed.

Figure 14.5 Iraklis, after his lesions healed, his sore feet were trimmed and he felt safe. (Image supplied by Agia Marina Donkey Rescue and reproduced with their permission.)

Ten-year-old Iraklis had lived alone his whole life, suffering from unrelenting physical pain and exhaustion and the emotional pain of loneliness so damaging to isolated donkeys. All this changed when he was finally able to move into the donkey field and meet his new donkey family. Despite a dust storm on the day, he was soon settled under the shade of a cool olive tree with the assorted family members taking it in turns to welcome Iraklis to his new life.

Iraklis was one of the lucky ones—when his owner blatantly beat him in public, it set in motion a team of animal advocates and carers who, working together, ensured Iraklis would never again suffer at the hands of humans. The few sanctuaries that can provide rescued donkeys with a dignified retirement struggle to meet the demand as the plight of working or abandoned donkeys becomes increasingly evident. Iraklis' story had a happy ending when he could finally display his happy, gentle nature and unique personality in the safety of Agia Marina Donkey Rescue, surrounded by donkey friends, but many more donkeys are waiting to be rescued.

REFERENCES

Agia Marina Donkey Rescue (2020). *Our story.* https://www.agia-marina-donkey-rescue.com/about

Bukhari, S.S.U.H., Parkes, R.S.V., Sneddon, L.U., & McElligott, A.G. (2024). The behavior and welfare of neglected species: Some examples from fish and mammals. *PeerJ, 12,* e17149. https://doi.org/10.7717/peerj.17149.

Clancy, C., Cooke, F., & Gorman, R. (2022). Learning with donkeys – A "more-than-human" approach to animal-assisted activities. *Society & Animals, 32*(4), 341–361. https://doi.org/10.1163/15685306-bja10103.

Clutton-Brock, J. (2012) *Animals as domesticates: A world view through history.* Michigan State University Press.

Dai, F., Dalla Costa, E., Burden, F., Judge, A., & Minero, M. (2019). The development of guidelines to improve dairy donkey management and welfare. *Italian Journal of Animal Science, 18*(1), 189–193. https://doi.org/10.1080/1828051X.2018.1503571

Farhat, S.F., McLean, A.K., & Mahmoud, H.F.F. (2020). Welfare assessment and identification of the associated risk factors compromising the welfare of working donkeys (*Equus asinus*) in Egyptian brick kilns. *Animals, 10*(9), 1611. https://doi.org/10.3390/ani10091611

González, F.J.N., Vidal, J.J., Jurado, J.M.L., McLean, A.K., & Bermejo, J.V.D. (2019). Dumb or smart asses? Donkey's (*Equus asinus*) cognitive capabilities share the heritability and variation patterns of human's (*Homo sapiens*) cognitive capabilities. *Journal of Veterinary Behavior, 33,* 63–74. https://doi.org/10.1016/j.jveb.2019.06.007

Goodrum, F., Theuri, S., Mutua, E. & Carder, G. (2022). The donkey skin trade: Challenges and opportunities for policy change. *Global Policy, 13,* 304–309. https://doi.org/10.1111/1758-5899.13072

Gross, M. (2022). A brief history of donkeys. *Current Biology, 32*(9), R985–R987. https://doi.org/10.1016/j.cub.2022.09.031

Lima, Y.F., Tatemoto, P., Reeves, E., Burden, F.A. & Santurtun, E. (2022). Donkey skin trade and its non-compliance with legislative framework. *Frontiers in Veterinary Science, 9,* 49193. https://doi.org/10.3389/fvets.2022.849193

Maśko, M., Wierzbicka, M., Zdrojkowski, Ł., Jasiński, T., Sikorska, U., Pawliń ski, B., & Domino, M. (2022). Comparison of donkey, pony, and horse dorsal profiles and head shapes using geometric morphometrics. *Animals, 12,* 931. https://doi.org/10.3390/ani12070931

Merkies, K., Paraschou, G., & McGreevy, P. D. (2020). Morphometric characteristics of the skull in horses and donkeys—A pilot study. *Animals, 10*(6), 1002. https://doi.org/10.3390/ani10061002

Milne, A.A., McCafferty, C., Kurtz, J. & Walt Disney Productions (1997). *Pooh's grand adventure: the search for Christopher Robin.* Grolier Book Club ed.

Murray, L.M.A., Byrne, K., & D'Eath, R.B. (2013). Pair-bonding and companion recognition in domestic donkeys, *Equus asinus, Applied Animal Behaviour Science, 143*(1), 67–74. https://doi.org/10.1016/j.applanim.2012.11.005

Nguyen, H.A.T., Guo, C., & Homberg, J.R. (2020). Cognitive bias under adverse and rewarding conditions: A systematic review of rodent studies. *Frontiers in behavioral neuroscience*, 14, 14. https://doi.org/10.3389/fnbeh.2020.00014

Norris, S.L., Little, H.A., Ryding, J., & Raw, Z. (2021). Global donkey and mule populations: Figures and trends. *PLoS One, 16*(2), e0247830. https://doi.org/10.1371/journal.pone.0247830

NSW Department of Primary Industries (2022). *NSW code of practice and standard operating procedures for the effective and humane management of feral donkeys.* https://www.dpi.nsw.gov.au/__data/assets/pdf_file/0006/1394655/Code-of-Practice-and-Standard-Operating-Procedures-for-the-Effective-and-Humane-Management-of-Feral-Donkeys.pdf

PETA (2024). *VICTORY: Africa bans killing donkeys for Chinese medicine.* https://investigations.peta.org/donkeys-slaughtered-for-skin-kenya/

PETA Australia (2023). *Suffering of donkeys and mules on Santorini continues – Help them now!* https://secure.peta.org.au/page/67585/action/1?locale=en-AU

Raspa, F., Cavallarin, L., McLean, A. K., Bergero, D., & Valle, E. (2019). A review of the appropriate nutrition welfare criteria of dairy donkeys: Nutritional requirements, farm management requirements and animal-based indicators. *Animals, 9*(6), 315. https://doi.org/10.3390/ani9060315

Seganfreddo, S., Fornasiero, D., De Santis, M., Mutinelli, F., Normando, S., & Contalbrigo, L. (2023). A pilot study on behavioural and physiological indicators of emotions in donkeys. *Animals, 13*(9), 1466. https://doi.org/10.3390/ani13091466

Shah, S.Z.A., Nawaz, Z., Nawaz, S., Carder, G., Ali, M., Soomro, N., & Compston, P.C. (2019). The role and welfare of cart donkeys used in waste management in Karachi, Pakistan. *Animals, 9*(4), 159. https://doi.org/10.3390/ani9040159

The Book of Pooh (2001). Season 1, Episode 16, Beastly Burden. *Walt Disney Home Entertainment.* https://www.imdb.com/title/tt5601890/

The Flicka Foundation Donkey Sanctuary (n.d). *About us.* https://www.flickafoundation.org.uk/about-us.html

Thiemann, A. & Foxcroft, A. (2016). Working across Europe to improve donkey welfare. *Veterinary Record*, 179(12), 298–300. https://doi.org/10.1136/vr.i4112

Tully, P.A.G. & Carr, N. (2020). The oppression of donkeys in seaside tourism. *International Journal of the Sociology Leisure, 3*, 53–70. https://doi.org/10.1007/s41978-019-00047-x

World Population Review (2025). *Donkey population by country 2025*. https://worldpopulationreview.com/country-rankings/donkey-population-by-country

15

Rabbits: Sensitive and curious companions

INTRODUCTION

There is no doubt that rabbits are cute. Soft toys are made in their image, and they have gained the status of a small, low maintenance "starter pet" for children, an assumption that has contributed to the poor treatment or abandonment of countless rabbits. For children, perceptions of rabbits as toylike animals have been reinforced by cartoon characters such as English author Beatrix Potter's *Peter Rabbit*, the white rabbit in *Alice in Wonderland* and Looney Tunes' *Bugs Bunny*. In some cultures, rabbits comprise part of the Christian Easter celebrations as an edible chocolate replica and symbol of a new beginning.

However, rabbits are also the fourth most farmed animal in the world, with many spending their short lives in cages with just a drinker and feeder and a wire mesh floor (Compassion in World Farming, 2025). Rabbit meat was once considered food of the poor, but in some western nations, it has become a gourmet food in restaurants and part of the urban farming movement (DeMello, 2014). Described as a superfood that is low in fat, nutritious and tasty, backyard rabbits risk being raised in a small hutch for a few months prior to slaughter by unregulated methods. Rabbit fur is also the only part of the global fur industry that has continued to grow in the face of consumer backlash against the more high-end furs. Rabbit fur is considered a "fun fur", used as a cheap trim on clothing and toys and marketed towards younger consumers. Less well known is that this fur comes from rabbits living on massive fur farms in abominable conditions. Their fur is not a by-product of the meat industry. Fur rabbits are raised specifically to be slaughtered at around six months, whereas meat rabbits are slaughtered at around 12 weeks.

Rabbit meat production data is not available for all countries, but some estimates suggest around 570 million rabbits (and hares) were slaughtered globally in 2021 (Compassion in World Farming, 2024). China produces more than half of global rabbit meat production, with Africa, other regions of Asia, the

DOI: 10.1201/9781003518853-16

Americas and Europe, especially Italy, Spain and France, accounting for most of the remainder (Cullere & Zotte, 2018).

Domestic rabbits can breed rapidly, leaving them open to genetic manipulations and rapid replacement of "stock". As a result, they now occupy many spaces in a human-dominated world: as micro-livestock for meat or fur, research subjects and companion animals, all of which have the potential to do harm to these sentient animals.

SOME BACKGROUND HISTORY

The integration of traditional archaeological approaches with genetic methods for explaining domestication trajectories has caused some discrepancies between the two techniques when tracing the history of rabbits (Irving-Pease et al., 2018). Roman scholar Marcus Terentius Varro (born 116 BCE, died 27 BCE) was one of the first to document breeding rabbits after the Romans encountered them on the Iberian Peninsula. Describing them as similar to the Spanish native Hare but with short legs, he instructed his wife to keep rabbits in a type of medieval warren (leporarium) and fatten them in hutches prior to slaughter (Clutton-Brock, 2012).

Archaeological evidence indicates that rabbits were hunted for fur and meat during the Epipaleolithic, Mesolithic and early Neolithic (10,000 BCE to 2,000 BCE) in the Iberian Peninsula and southwest France, but it was not until the Middle Ages (fifth to fifteenth centuries) that they were intentionally transported to Europe in the start of the domestication process. These rabbits were very similar to their wild relatives, with skeletal changes not appearing until almost 2,000 years after early historical accounts of their use in captivity. These skeletal changes coincided with evidence of rabbits being kept as companion animals rather than just for meat and fur, a precursor of the many uses that humans would subsequently impose on domesticated rabbits (Irving-Pease et al., 2018). Cumulative evidence suggested that the domestication of rabbits occurred along a continuum dependent on their uses at any one point in history, culminating in selective breeding for size, temperament, colour and facial features (for example, flat-faced or lop eared) to match their use as micro-livestock, research subjects or companion animals (DeMello, 2014).

Intensive rabbit farming for meat production arose in parts of the USA in the 1920s and in several European countries, including Italy, France, Spain and Belgium by the 1970s (Cullere & Zotte, 2018). Artificial insemination, development of highly productive hybrid strains and increased knowledge of nutritional requirements and disease control prompted larger and larger indoor facilities to produce rabbit meat at relatively low prices. However, this was accompanied by mounting criticism from animal advocacy groups and consumers about the welfare and ethical issues of inconsistently regulated intensive systems.

RABBITS IN CONTEMPORARY SOCIETY

As micro-livestock, rabbits offered economic benefits in terms of space and food requirements compared to larger farmed animals. They required less food to produce meat with a higher ratio of protein, energy, calcium and vitamins, coupled with reduced amounts of cholesterol, fat and sodium compared to other meats. They grew fast, reproduced rapidly, were unencumbered by any religious restrictions related to their consumption and their pest status in many countries attracted less welfare considerations (Walia & Kaur, 2023). They presented as the almost perfect commodified animal with the ability to transform around 20 per cent of the protein they consumed into muscle mass, close to the poultry conversion rate but higher than beef (8–12 per cent) and pigs (16–18 per cent). This ratio, a key requirement for predicting economic performance and environmental sustainability in the farming system, made the rabbit a very lucrative farming option. However, the ambiguous space that rabbits occupied as both food and companion animal saw some consumers still retaining an aversion to eating look-alikes of Beatrix Potter's famous rabbit characters, Peter Rabbit or his sisters Flopsy, Mopsy and Cotton-tail.

Researchers Fabio Napolitano and colleagues (2010) identified a range of consumer trends that influenced the acceptance of animal-based products, including: a reliance on intrinsic cues (the visible characteristics of the product, such as colour and appearance of meat); extrinsic cues (brand, label, price, origin, production and nutritional information) and credence characteristics such as healthiness and ethical process characteristics, encompassing the farming practice and related animal welfare aspects. Credence characteristics were not evident before or after purchase but had become increasingly important to consumers. Within this framework, knowledge that cheap rabbit meat was largely derived from animals living in filth and crammed into confined wire mesh cages was an inevitable source of cognitive dissonance especially when juxtaposed against rabbits as companion animals. Living in cages fitted with wire floors caused leg, footpad and paw injuries, while environmental stressors associated with temperature fluctuations, ammonia build-up, noise and crowding increased frustration, stress and psychological health issues. In confined settings, stereotypies proliferated, including biting cage corner bars, excessive body grooming, frequent shaking of the head and moving around in repeated circles (Voiceless, 2020).

Animal advocacy and consumer criticism prompted some intensive farms around the world to implement "enriched cages" to provide more space and with additions such as gnawing sticks. However, rabbits were still denied natural behavioural needs of socialising within a chosen group, playing, nest building, hiding, exploring and foraging, and they still died well before their natural lifespan. Collective outdoor pens have been posited as an improved welfare option but unless designed from a rabbit's perspective, they can impose dangers to some rabbits (Siddiqui et al., 2023). Raising rabbits in large groups where they can socialise and interact sounds ideal but if groupings are forced, conflict between socialisation and territoriality can arise. In the wild, there is usually a gender-separated hierarchy, with ranking fights ending when the inferior rabbit retreats and hides. Spatial restrictions in outdoor enclosures without sufficient retreating distances meant aggressive confrontations could inflict injury or death. Without an understanding and consideration of each rabbit's social, spatial and safety needs, there was a risk that only the more dominant members of the group would benefit.

Irrespective of farming methods, producers have manipulated the rabbit's ability to suckle and be pregnant at the same time to achieve higher yields. As a prey species that cannot fight predators, maximum reproduction is a species survival technique. Doe rabbits enter a state of postpartum oestrus immediately following delivery of their young (González-Mariscal et al., 2022). During this time, they can ovulate, mate and become pregnant again, resulting in a concurrency of pregnancy and lactation that accelerates and maximises reproduction. This has a downside for rabbit breeders when there is a decline in milk production from around day 20 of the doe's 31-day pregnancy, and consequently, a lower body weight gain in the unweaned youngsters and increase in production costs. Some rabbit farms globally have introduced a process called "biostimulation" to address this reduction in growth rate and maximise productivity.

Separation of the mother and infants for 48 hours around lactation days nine to 12 does not significantly affect the infants' weight at weaning and allows the doe to be mated or undergo artificial insemination (González-Mariscal et al., 2022). This second pregnancy is concurrent, but importantly not aligned with the first lactation, and avoids any decline in body weight of unweaned youngsters. They remain on track to be weaned at one month and slaughtered at two months with maximum body weight. Rodent studies have identified some negative impacts on an infant's developing brain following mother–young separation in early lactation, including reproductive behaviour, neuroendocrine reactivity to stress, learning and memory. Replicated studies among rabbits reported almost double the level of corticosterone in response to mildly aversive stimulation among young rabbits separated for 48 hours compared to control rabbits. Importantly, the increased corticosterone secretion in the early neonatal period affected their responses to mildly aversive stimulation into adulthood, suggesting that young rabbits need a full complement of maternal care to allow normal development of some components of the stress-reactivity system. The increased production and economic benefits of biostimulation potentially came with long-lasting complex consequences for the rabbits. Selectively breeding to produce larger litter sizes compounded this issue, putting the mother's body under increased strain and risking loss of body condition, metabolic diseases and spinal deformity. While these factors reduced their breeding life, they were quickly replaced with the youngsters they were forced to churn out.

Rabbits in research

The rapidity with which rabbits reproduce also made them cheap and plentiful research animals. Selectively bred to be docile and gentle, they yield a larger amount of sample material compared to rats and mice and have become popular models for conditions including cardiovascular disease, intestinal immunity, reproduction, lupus, arthritis, cancer, Alzheimer's disease as well as a range of infectious diseases. In 1908, a Russian physician fed rabbits milk, meat and eggs to understand aortic atherosclerosis, paving the way for rabbit models in human cardiovascular disease and lipid metabolism in preference to rats and mice (Fan et al., 2021). By the 1980s, the development of genetically modified rabbits tailored to individual research projects was underway, with the first use of transgenic rabbits in the field of cardiovascular disease adopted in 1994.

Rabbits played a role in Nobel Prize winner Arvid Carlsson's work around the role of dopamine in Parkinson's Disease, the development of the rabies vaccine by Louis Pasteur in 1881 and pioneering studies on syphilis (Esteves et al., 2018). The identification of white–grey matter ratios similar to the human brain, as well as the similar timing of the perinatal brain white matter maturation, consolidated their use in neuroscience, with rabbit pups providing valuable information on perinatal neurodevelopment, especially in relation to chemosensory processes and social and feeding behaviour (Schneider et al., 2018).

The controversial and distressing Draize eye test, developed in 1944, favours albino rabbits to assess if a substance would cause eye irritation to humans (New Zealand Anti-Vivisection Society, 2023). Iconic images of trapped rabbits with weeping, red eyes became a rallying point for the horrors of animal treatment in the name of science and commercial industry.

Companion animal rabbits

Those rabbits destined to become companion animals rather than sources of meat, fur or research models should have enjoyed a more privileged lifestyle. However, a brief perusal of animal shelters and farmed animal sanctuary residents debunks this popular misconception. Rabbits experience relentless and thoughtless abandonment, especially at certain times of the year. Newspaper headlines such as "More domestic bunnies on the loose: Rabbit dumping picks up in the summer after Easter gifts go stale" (Rhone, 2017) and "Rescue groups see rise in rabbit abandonment" (Bain, 2007) make it clear that even the elite in the bunny world are not immune to human cruelty.

People for the Ethical Treatment of Animals (PETA, 2025) describe this as "bunny fever" in a poignant reminder that rabbits are sentient animals not bred to survive alone: "As sure as the sun rises every Easter morning, many moms and dads give in to 'Easter bunny' temptation and buy a rabbit for their kids, vastly underestimating the amount of care that bunnies require. Weeks later, when 'bunny fever' has subsided, many will consign now-unwanted bunnies to outdoor hutches, dump them at animal shelters, or simply set them free outdoors, where they will starve or be killed by predators".

It is neither difficult nor expensive to purchase a rabbit, especially if using online classified advertising platforms. A United Kingdom study estimated around a quarter of a million rabbits each year were purposefully bred in United Kingdom alone for the online pet sales market, with many sold unvaccinated, unneutered and with the potential for health issues associated with the popular lop-eared and brachycephalic breeds (Gosling et al., 2018).

There are at least nine breeds with the lop-eared heritable trait and associated altered skull morphology (Johnson & Burn, 2019). These rabbits can be at greater risk of ear infection, such as otitis externa, as identified among dog breeds with pendulous ears. The altered skull morphology creates an additional welfare concern shared by brachycephalic and dwarf breeds: congenital malocclusion when shortening of the upper jaw misaligns upper and lower teeth and promotes overgrowth. Without regular care, rabbits experience pain and suffering through lacerations of the mouth, abscesses and prolific tears or pus running down their face from misshapen tear ducts. A study among rabbit breeds with "short faces" compared to "long faces" reported that short faces were three times more likely to be diagnosed with dental disease (Jackson et al., 2024). When neglected, the pain and physical restrictions imposed by malocclusion can result in death. Veterinary care at this late stage can be expensive prompting relinquishment or abandonment of the animal that was meant to be low maintenance.

Selective breeding for the "cute factor" associated with the shortened, broadened paedomorphic features of brachycephalic and lop-eared rabbits has seen the Lion Head, Netherland Dwarf and Miniature Lop breeds become irresistibly popular worldwide. For example, some estimates suggest that the Miniature Lop and Netherland Dwarf comprise over half the rabbit population in United Kingdom (Campbell, 2024). In some European countries, the rabbit has been ranked as the third most popular companion animal (Skovlund et al., 2023), reinforced by the common perception of rabbits as low investment "starter pets" for children. While there is no doubt that many are beloved and cherished family members, some end up living in environmental conditions ranging from restricted space allowance, solitary housing, imposed group formations or with limited behavioural opportunities. As long as demand persists for these rabbits with their emotive faces, breeders will continue to produce them with resultant physical and psychological welfare implications when bred in less-than-ideal circumstances, as Juniper's story highlights.

Freedom and a new friend—Juniper and Theo, Clement Rabbit Rescue, Wales

Juniper's emotive brown eyes were striking. When she first arrived at Clement Rabbit Rescue in North Wales, they held fear and trepidation but gradually this was replaced with friendship, trust and security.

Like many animal rescues, Clement Rabbit Rescue began with one signature animal who was the catalyst to provide a safe haven for many more neglected and abandoned rabbits (Clement Rabbit Rescue, 2024–2025). For founder Paige, this was Bertie, sole survivor of a metal crate full of emaciated, urine and faeces-soaked baby rabbits delivered to a pet shop to sell. By 2019, Clement Rabbit Rescue had begun rescuing, rehabilitating and caring for rabbits, many with lifelong health problems, while simultaneously educating and spreading awareness of the needs of these loving and sensitive animals. More than three hundred rabbits later, the demand remained unabated as rabbits were subjected to the brutality of the breeding trade, neglect or abandonment. Caring for neglected animals can bring grief and loss for Paige and the remaining rabbits, but also the knowledge that for a moment in time, these rabbits were free of human cruelties.

During Juniper's time at Clement Rabbit Rescue, she not only experienced freedom after a lifetime of relentless breeding, but she also found a friend with whom to share her new life. Juniper had lived among several unneutered males, subsequently mothering countless litters while in a state of constant terror. She was covered in open wounds, evidence of the attacks from other rabbits as they all struggled to survive locked in a backyard, drinking from muddy puddles and with no shelter. Inside the house, young rabbits were crammed in filthy cages caked in urine and faeces.

Juniper struggled to maintain weight, and numerous tests could not determine a definitive diagnosis. However, her bright eyes belied any physical frailty as she remained active, inquisitive and loving. Despite her traumatic start to life, Juniper became one of Paige's kindest, most gentle and affectionate residents, a testament to the forgiving nature inherent in so many animals once they feel safe. This was never more evident than when Juniper met Theodore (aka Theo).

Several months after Juniper's arrival, Theo was found alone in a hutch at the side of the road. Dehydrated and with long matted fur, Theo's multiple medical conditions confirmed that he was not a young man. Severe eye injuries and the possibility of losing his sight altogether, respiratory related issues and a testicular tumour the size of a golf ball signified long-term suffering. Further investigations revealed severe malocclusion, corneal ulcers, maxillary sinusitis, arthritis and parasites—the list went on and on for this gentle old man who nevertheless demonstrated he was not yet ready to give up on life.

Meeting Juniper may also have given him motivation to live. The two became best friends in a short time, each seeming to crave the safety and security of this relationship in their new home. Having Juniper around to guide him, groom him, keep his eyes clean and provide companionship made a world of difference physically and emotionally to Theo.

As prey animals, rabbits evolved to rely on the company of their group members to warn them of predators, but also to keep warm and to just hang out. In a study reminiscent of American psychologist Harry Harlow's (1958) seminal findings that infant rhesus monkeys sought comfort from a cloth covered surrogate mother in preference to milk, researchers Shirley Seaman and colleagues (2008) demonstrated the importance of a friend among laboratory rabbits. The researchers reported that single-housed female laboratory rabbits showed a level of motivation for social contact similar to that for food, highlighting the ethical implications of keeping any rabbits in isolation.

Worldwide, hundreds of thousands of nameless rabbits live and die with no acknowledgement of their existence. When Juniper arrived at Clement Rabbit Rescue, she became someone—with a name, a personality and friends—and her subsequent death was mourned by Paige, Theo and all who had known this loving and unique rabbit.

Figure 15.1 Juniper's emotive brown eyes reveal her gentle and affectionate nature. (Image supplied by Clement Rabbit Rescue and reproduced with their permission.)

Figure 15.2 Theo, in the middle, receiving comfort from Daisy and Baxter following the loss of Juniper. (Image supplied by Clement Rabbit Rescue and reproduced with their permission.)

FEELINGS AND EMOTIONS

A range of studies have confirmed that the personalities of domesticated animals differ from their wild relatives and that the resultant behavioural differences can affect the development of the nervous system. Domesticated rabbits are no exception, demonstrating a reduced brain volume relative to body size compared to their wild ancestors (Brusini et al., 2018). Researchers Irene Brusini and colleagues investigated this trend, and with the assistance of high-resolution MRI, proposed that this reduction was not uniform across all parts of the brain. Instead, they identified specific areas of variance that could affect both personality and behaviour. Of particular note was a reduction in size of the amygdala and enlargement of the medial prefrontal cortex in domesticated rabbits compared to wild rabbits. Both were linked to fear conditioning and fear reduction changes. The amygdala is the brain structure associated with processing emotional memory and triggering the fight-or-flight response, a key survival strategy. The medial prefrontal cortex supports the processing of social information and emotional control, with enlargement of this area linked to fear reduction through safety learning and emotional control. Selective breeding for tame, non-fearful, non-attentive and non-aggressive rabbits compliant with various human needs was therefore visible in their very brain structure, exacerbating the risk that domesticated rabbits experienced when abandoned or released in the wild. Domesticated rabbits no longer had the ability to cope alone and avoid predation.

Domestication did not, however, change other behaviours and attributes of rabbits. They remain sensitive, curious and intelligent animals who require exercise, mental stimulation and the ability to express natural behaviours such as foraging for food, grooming and socialising with their friends. When a lack of space or environmental enrichment—whether in intensive farm settings, research facilities or inappropriate backyard settings—prevents the expression of needs and emotions, the resultant stress build-up can cause psychological problems that manifest in stereotypies or boredom, apathy and depression. This perpetuates the stereotype of rabbits as lacking personalities, emotions and intelligence, when nothing could be further from the truth. This perception is compounded by the rabbit's natural stoicism and reluctance to show pain. Rabbits do not vocalise much, but when they do scream, their suffering has reached an unbearable level.

Attributing emotional capacity to an animal is significantly associated with more positive treatment, a finding supported among a study of over 1,500 rabbit owners in the United Kingdom (McMahon & Wigham, 2020). The researchers concluded that to improve the welfare of this often-overlooked companion animal, there needed to be a targeted approach to improving perceptions of their sentience. Perceiving rabbits as sentient, sociable and complex animals would trigger empathy and compassion as opposed to seeing them as low maintenance, disposable starter pets.

Spend some time with domesticated rabbits who feel safe and secure, and they will play, run and spin just for the fun of it, all behaviours that their wild relatives perfected so as to escape predators (DeMello, 2014). They will also develop lasting bonds that demonstrate their emotional capacity when not forced to meet human needs.

Spring and Snow, lasting bonds—Flow Farm and Sanctuary, China

The sisters Spring and Snow were inseparable. For almost eight years, they foraged together, cuddled together and groomed each other at Flow Farm and Sanctuary in Zhongshan, Guangdong Province, China. Founder Justine Kwok and her team respected all residents as individuals, allowing each one to express their unique personalities and preferences in safety, and Spring and Snow thrived in this environment.

When Spring started showing signs of weakness in one of her hind legs, she received two weeks of intensive care at the vet clinic to determine a diagnosis. Rabbits can be predisposed to vertebral fractures due to their relatively light bones—only seven to eight per cent of their body weight (Carpenter, 2006)—but this was not the case for Spring. The vet determined Spring was suffering from a neurological disease, another common condition among rabbits who live longer than the two months allotted to them in intensive farm systems.

After two weeks' separation, Snow greeted the return of her friend with a two-hour intensive cleaning session. Presumably seeking to remove the foreign smell of the vet clinic, it also demonstrated the strong bond between the two sisters and reinforced the need to maintain Spring's quality of life for as long as possible. However, even with medication, Spring's condition deteriorated until she could no longer use her hind leg.

With one functional hind leg, Spring could still drag herself around the rabbit enclosure, and Snow was always nearby to help out. When the team noticed hay wrapped around the ankle of Spring's functional leg, they assumed the pieces had become entangled as she crawled around and so removed the "hay bracelet". The next day more hay was wrapped around Spring's lower leg, a trend that continued each time it was removed. The intricacy of the hay bracelet bemused the team who had no idea why Spring persisted in using hay in this way.

When investigating behaviours among growing rabbits, researchers Katalin Szendrő and colleagues (2015) reported that hay provided a useful environmental enrichment. Rabbits given hay interacted more with the hay than rabbits given gnawing sticks, and importantly, rabbits in cages enriched with hay also showed less abnormal behaviours. While these findings demonstrate the comfort derived from access to hay and may give some insight into Spring's determination to maintain a hay braid around her functional leg, nobody will ever know for sure. However, the team recognised and respected Spring's preference to maintain this routine, and she continued creating these hay bracelets for reasons known only to herself.

The neurological condition progressed, eventually leaving Spring paralysed in both legs. A custom-made wheelchair soon had her wheeling herself around the garden as she foraged for food, with Snow always by her side. Spring lived another six months after her original diagnosis, and during that time, Snow continued to nurture, groom and cuddle her paralysed sister.

When neuroscientist and psychobiologist Jak Panksepp mapped out the foundations of human and animal emotions, he identified oxytocin and prolactin (neurochemicals) as major social bonding pathways in the mammalian brain (Panksepp & Watt, 2011). The loss of an important attachment figure diminished these neurochemicals, resulting in a sense of withdrawal and feelings of separation distress. In humans, this process would be called grief and linked to the capacity to form and lose bonds, something that is also evident in a range of animals. Their grief may not manifest in the same expression across species, but that does not mean it is not present. Anthropologist Barbara King (2013) identified two factors that increased the likelihood of an animal experiencing grief. Firstly, the two animals had chosen to spend time together beyond survival-oriented behaviours such as hunting and mating; and secondly, at the death of one animal, the other animal changed their behavioural routine (for example, bodily posture, appetite, failure to thrive). When these changes in behaviour extended for a prolonged period and were accompanied by appearances of distress, it was likely that the animal was feeling grief.

As prey animals, rabbits are accustomed to hiding their feelings, but to the team who had known these bonded sisters for eight years, Snow seemed sad and her spirit diminished. She lost her appetite, and four months after Spring's death, Snow passed away.

Figure 15.3 Sisters Spring and Snow were inseparable. (Image supplied by Flow Farm and Sanctuary and reproduced with their permission.)

Figure 15.4 Spring's custom-made wheelchair allowed her to wheel herself around the garden. (Image supplied by Flow Farm and Sanctuary and reproduced with their permission.)

ETHICAL IMPLICATIONS

Compared to other farmed and companion animals, rabbits occupy an ambiguous space with limited protections in both domains. Caged layer hens have received negative publicity about conditions in intensive farm settings, raising consumer ire and making it increasingly difficult to rationalise away the cognitive dissonance. Caged rabbits endure similar living conditions to caged hens, but perhaps because their flesh is not as widely in demand as eggs, they have received less acknowledgement of their suffering. This is changing as more people recognise them as sentient beings rather than production units, but the disconnection between industrial agriculture and consumption of the end product continues to make it difficult to ensure long-term that no suffering will occur in these settings.

When animal activist Tom Regan wrote the preface to the 2004 edition of his seminal book *The Case for Animal Rights*, he expressed dissatisfaction when hog producers or fur mill operators made changes to treat animals more humanely. Instead, he demanded an end to raising animals for food and killing them for their fur with the statement, "Not larger cages", we declare, "empty cages" (p. xiv). Decades later, rabbits remain in cages and slaughtered for their meat and their fur, but their versatility also sees them suffering in laboratory settings, as selectively bred pets or when abandoned. Common stereotypes ranging from "breeding like rabbits" to "starter pet" provide ample strategies to quell the cognitive dissonance about using and abusing these sensitive animals.

REFERENCES

Bain, B. (2007, Sep 28). Rescue groups see rise in rabbit abandonment: Rescue groups see rise in rabbit abandonment. *McClatchy - Tribune Business News* Retrieved from https://www.proquest.com/wire-feeds/rescue-groups-see-rise-rabbit-abandonment/docview/462235054/se-2

Brusini, I., Carneiro, M., Wang, C., Rubin, C.J., Ring, H., Afonso, S., Blanco-Aguiar, J.A., Ferrand, N., Rafati, N., Villafuerte, R., & Smedby, Ö. (2018). Changes in brain architecture are consistent with altered fear processing in domestic rabbits. *Proceedings of the National Academy of Sciences of the United States of America, 115*(28), 7380–7385. https://doi.org/10.1073/pnas.1801024115. PMID: 29941556; PMCID: PMC6048517

Campbell, H. (2024). Anatomical evaluation of brachycephalism in cats and rabbits. *The Veterinary Nurse, 15*(4). https://www.magonlinelibrary.com/toc/vetn/15/4; https://doi.org/10.12968/vetn.2023.0004

Carpenter, J.W. (2006). Diagnosing and treating common neurological diseases in rabbits. *Veterinary Medicine*, November 2006, 728–740.

Clement Rabbit Rescue (2024–2025). *Who's behind the rescue*? https://www.clementrabbitrescue.com/about

Clutton-Brock, J. (2012) *Animals as domesticates: A world view through history*. Michigan State University Press.

Compassion in World Farming (2024). *Information Sheet 1 – Rabbit meat production – Global*. https://www.compassioninfoodbusiness.com/resources/rabbits/rabbit-meat-production-global-review-2024/

Compassion in World Farming (2025). *Rabbit welfare*. https://www.ciwf.eu/farm-animals/rabbits/rabbit-welfare/

Cullere, M. & Dalle Zotte, A. (2018). Rabbit meat production and consumption: State of knowledge and future perspectives. *Meat Science, 143*, 137–146. https://doi.org/10.1016/j.meatsci.2018.04.029

DeMello, M. (2014). Rabbits in captivity. In L. Gruen (Ed.), *The ethics of captivity* (pp. 77–89). Oxford University Press. https://doi.org/10.1093/acprof:oso/9780199977994.003.0006

Esteves, P.J., Abrantes, J., Baldauf, H.M., BenMohamed, L., Chen, Y., Christensen, N., González-Gallego, J., Giacani, L., Hu, J., Kaplan, G., & Keppler, O.T. (2018). The wide utility of rabbits as models of human diseases. *Experimental & Molecular Medicine, 50*, 1–10. https://doi.org/10.1038/s12276-018-0094-1

Fan, J., Wang, Y., & Chen, Y.E. (2021). Genetically modified rabbits for cardiovascular research. *Frontiers in Genetics, 12*, 614379. https://doi.org/10.3389/fgene.2021.614379

González-Mariscal, G., Hoy, S., & Hoffman, K.L. (2022). Rabbit maternal behavior: A perspective from behavioral neuroendocrinology, animal production, and psychobiology. In G. González-Mariscal (Ed.), *Patterns of parental behavior. Advances in neurobiology, vol* 27 (pp. 131–176). Springer. https://doi.org/10.1007/978-3-030-97762-7_5

Gosling, E.M., Vázquez-Diosdado, J.A., & Harvey, N.D. (2018). The status of pet rabbit breeding and online sales in the UK: A glimpse into an otherwise elusive industry. *Animals, 8*(11), 199. https://doi.org/10.3390/ani8110199

Harlow, H.F. (1958). The nature of love. *American Psychologist, 13,* 673–685.

Irving-Pease, E., Frantz, L., Sykes, N., Callou, C., & Larson, G. (2018). Rabbits and the specious origins of domestication. *Trends in Ecology and Evolution, 33*(3), 149–152.

Johnson, J.C. & Burn, C.C. (2019). Lop-eared rabbits have more aural and dental problems than erect-eared rabbits: A rescue population study. *Veterinary Record, 185*(24), 758. https://doi.org/10.1136/vr.105163

King, B.J. (2013). *How animals grieve.* University of Chicago Press.

McMahon, S.A., & Wigham, E. (2020). 'All ears': A questionnaire of 1516 owner perceptions of the mental abilities of pet rabbits, subsequent resource provision, and the effect on welfare. *Animals, 10*(10), 1730. https://doi.org/10.3390/ani10101730

Napolitano, F., Girolami, A., & Braghieri, A. (2010). Consumer liking and willingness to pay for high welfare animal-based products. *Trends in Food Science & Technology, 21*(11), 537–543. https://doi.org/10.1016/j.tifs.2010.07.012

New Zealand Anti-Vivisection Society (2023). *Rabbits.* https://www.nzavs.org.nz/case-studies/rabbits

Panksepp, J. & Watt, D. (2011). What is basic about basic emotions? Lasting lessons from affective neuroscience. *Emotion Review, 3*(4), 1–10. https://doi.org/10.1177/1754073911410741

PETA (2025). *Bunny fever: Don't catch it!* https://www.peta.org/features/bunny-fever/

Rhone, N. (2017, August 27). PETS: More domestic bunnies on the loose: Rabbit dumping picks up in the summer after Easter gifts go stale. *Atlanta Journal-Constitution* [Atlanta, GA], p. S2.

Schneider, N.Y., Datiche, F., & Coureaud, G. (2018). Brain anatomy of the 4-day-old European rabbit. *Journal of Anatomy, 232*(5), 747–767. https://doi.org/10.1111/joa.12789

Seaman, S.C., Waran, N.K., Mason, G., & D'Eath, R.B. (2008). Animal economics: Assessing the motivation of female laboratory rabbits to reach a platform, social contact and food. *Animal Behaviour, 75*(1), 31–42. https://doi.org/10.1016/j.anbehav.2006.09.031

Siddiqui, S.A., Gerini, F., Ikram, A., Saeed, F., Feng, X., & Chen, Y. (2023). Rabbit meat—Production, consumption and consumers' attitudes and behavior. *Sustainability, 15*(3). https://doi.org/10.3390/su15032008

Skovlund, C.R., Forkman, B., Lund, T.B., Mistry, B.G., Nielsen, S.S. & Sandøe, P. (2023). Perceptions of the rabbit as a low investment 'starter pet' lead to negative impacts on its welfare: Results of two Danish surveys. *Animal Welfare, 32,* e45, 1–13. https://doi.org/10.1017/awf.2023.41

Szendrő, K., Szendrő, Zs., Matics, Zs., Dalle Zotte, A., Odermatt, M., Radnai, I., & Gerencsér, Zs. (2015). Effect of genotype, housing system and hay supplementation on performance and ear lesions of growing rabbits. *Livestock Science, 174,* 105–112. https://doi.org/10.1016/j.livsci.2015.01.008

Voiceless (2020). *Factory farming of domestic rabbits for meat in Australia.* https://voiceless.org.au/factory-farming-of-domestic-rabbits-for-meat-in-australia/

Walia, S.S. & Kaur, T. (2023). *Basics of integrated farming systems.* Springer Singapore. https://doi.org/10.1007/978-981-99-6556-4

Conclusion: Crossroads in animal welfare

It is hard to ignore or deny that farmed animals have the capacity to feel what is happening to them. They can feel fear, pain, frustration, anger and sadness—but they can also feel pleasure, excitement and love when in a safe environment. Philosophers, animal advocates, neuroscientists, biologists and ethologists have all made this point in various ways over the years. Ruth Harrison, Tom Regan, Peter Singer, Jane Goodall, Carl Safina, Jonathan Birch, Melanie Challenger, Marc Bekoff—the list goes on and on—and yet animals continue to be cruelly and brutally used, abused and slaughtered globally.

The invisibility of intensive animal agriculture, economic demands for cost-efficiency and end products that look nothing like the source animals contribute in some way to the persistence of these cruelties. However, the other major obstacles are cognitive dissonance and the strategies people enlist to mitigate this discomfort. Cognitive dissonance is unpleasant, but for many people, the pleasure and normalisation of meat eating and the use of animal products outweigh any fleeting guilt or uncertainty. Improvements to animal welfare have been made over the decades since Harrison's *Animal Machines* and actions stemming from this disclosure, but there is still a long way to go.

Farmed animals are often forced to suffer in silence, and it is only when they enter a farmed animal sanctuary that they are given a voice and a choice in their lives. Marc Bekoff (2024) commented that if we are to maintain any degree of hope for the future, it is important to focus on the successes being made for nonhuman animals, not just the challenges and harms still being done to them. The stories in this book speak of new lives for the diverse animals from the 23 farmed animal sanctuaries, thus highlighting some of the individual successes in the face of ongoing systemic animal cruelty.

In 2019, photographer Isa Leshko published a book entitled *Allowed to Grow Old. Portraits of Elderly Animals from Farm Sanctuaries*. This book drew attention to another success in farmed animal sanctuaries—animals allowed to age with dignity and on their own terms, an impossibility in intensive animal agriculture. Not all the animals featured in the preceding chapters are old, but they demonstrate wisdom and dignity in the way they have approached their new life. Irrespective of previous physical and psychological harms, they appear determined to live their lives to the fullest and on their terms. Farmed animal sanctuaries can only home a minute proportion of animals born into, or trapped within, a system that benefits only humans, but they can also fulfil a crucial role by putting a face and a voice—albeit a human voice—to some unique individuals. This provides an opportunity for people to recognise similarities between these

DOI: 10.1201/9781003518853-17

sentient animals and themselves, thus stimulating empathy rather than the continuation of destructive strategies to counter cognitive dissonance. Empathy requires a person to imagine how they would feel if they had to endure the same conditions from which an animal escaped, and while this may initially be painful, it can provide a first step towards change.

Domestication changed animals, and yet it will be humans who need to change if some of the long-term atrocities are to be corrected. For decades, humans have been modifying domestic animals to make them more amenable to human needs, sometimes with disastrous consequences for the animal, the species and the environment. Taken to extremes, the concept of psychological disenhancement proposed a means to make domestic animals more suited to the environments of industrialised farming by reducing their capacity to experience negative affective states. However, this also meant reducing capacity for positive affective states and the creation of animals with limited perceptual and/or cognitive capacities (Fischer, 2020).

Beilharz and Zeeb (1981) summed up the dilemma succinctly: "If we succeed in showing that animals are suffering under particular conditions of husbandry, what is to be done? There are two approaches. Either (1) we change the animals genetically and adapt them to the environment, or (2) we improve the environment" (p. 6).

Changing the animal genetically might turn a miserable life into a manageable life for an animal no longer able to recognise that they exist within an oppressive environment, but this fails to eliminate the root cause of suffering—an exploitive food production system for human gain. Farmed animal sanctuaries, on the contrary, choose to change the environment for as many animals as possible to counter some of the global suffering.

While the mammals and birds featured in this book remain at risk of ongoing ethical and welfare issues, there are billions of other animals even more hidden from the public gaze and vulnerable to unimaginable cruelties. Drawing public attention to calves, lambs, kids, chicks, ducklings and other mammals and birds with emotive eyes or paedomorphic features can present a relatively easier task than seeking compassion for some of the less "cuddly" animals—the reptiles, amphibians, fish and invertebrate animals such as shrimps and insects—also confined in ways contrary to their natural behaviours.

Crocodiles have been around for 100 million years, display unique personalities and hold great cultural significance for Australia's Aboriginal and Torres Strait Islander peoples (RSPCA, 2025). However, every year, around 150,000 are slaughtered to provide skins for luxury bags and accessories in the high-end fashion industry. Instead of living in their natural environments for up to 70 years, crocodiles and their eggs are taken and forced to live in barren concrete pens or tight wire cages to protect the valuable skin of their underbelly. After two or three years, they are slaughtered and skinned in the name of fashion—or pet food. Crocodiles can experience pain, stress, fear and anxiety, but they lack the mammal and bird cute factor and are farmed in remote areas far from consumer sight.

Nile crocodiles in Zimbabwe are also confined to concrete pits for their entire life, while alligators in parts of the USA suffer a similar fate for their meat and hides (PETA, 2025). The first commercial alligator farm was opened in Florida in the 1890s, and by the 1980s, indoor controlled farming provided a loophole to avoid hunting alligators protected by the Endangered Species Act. Market-ready alligators could be "grown" in under two years, thus separating farming from hunting (Guarcello et al., 2023). Like crocodiles, alligators have the capacity to play and enjoy themselves when in their natural habitat. However, overcrowding in industrial farm settings introduces behaviours suggestive of fear, distress and anxiety through increased incidents of aggression and serious fighting among animals who normally co-habit peacefully (Learmonth, 2020).

Fish are another group that miss out on cuddly cuteness, and yet, global farmed fish—called aquaculture—totals around 124 billion annually (Darlim, 2024). Add this to the more than one and a half trillion fish caught in the wild each year, not including bycatch and discards, and it is clear that these sentient animals are well and truly commodified. Fish are able to process noxious stimuli centrally even without the neocortex found in mammals and show a preference for areas of a tank with pain relievers if hurting (Balcombe 2016). In aquaculture, thousands of fish can be crammed together into pools or sea pens, causing injuries and wearing down their fins from rubbing against the sides of tanks or nets or other fish. Behavioural anomalies reminiscent of land factory farmed animals' stereotypies include swimming in endless circles, suggestive of a strong negative psychological impact when unable to swim freely.

A 2016 study of growth-stunted farmed Atlantic salmon reported that compared with healthy fish, growth-stunted salmon recorded increased basal brain serotonergic activity and cortisol production indicative of a chronic activation of the brain serotonergic system (associated with regulating mood) and the hypothalamic-pituitary-interrenal axis system (associated with regulating stress responses) (Vindas et al., 2016). Unfortunately, fish welfare research is limited, and as with many other farmed animals, it tends to focus on maintaining fish health to maximise the quality of the end "product" rather than addressing negative implications for the fish themselves when trapped in the factory farm system.

Compounding welfare issues of farmed fish is the differences in standards in many different countries with aquaculture (Animal Charity Evaluators, 2020). Globally, China appears to have the largest fish farming by volume, predominantly of carps. Europe mainly farms salmonids, such as rainbow trout and Atlantic salmon, while the USA catfish aquaculture comprises around 80 per cent of the nation's farmed fishes by tonnage. Six countries produce 92 per cent of the world's farmed salmon: Norway, Chile, Canada, the United Kingdom, Australia and New Zealand. Researchers Singh and colleagues' (2024) analysis of salmon aquaculture mortality rates between 2012 and 2022 in these countries identified that the mass mortality events had increased in frequency during that period (Singh et al., 2024). Crocodile, alligator and fish numbers cannot compete with the more than 440 billion shrimp farmed annually. Hidden away from consumers, the treatment of shrimp, a sentient decapod crustacean, raises huge ethical concerns. In 2021, the sentience of both decapod crustaceans (crabs, lobsters, crayfish, shrimps and prawns) and cephalopod molluscs (squid, octopus and cuttlefish) was officially recognised when the United Kingdom House of Commons passed an amendment to the *Animal Welfare (Sentience) Bill* (2021) to include these animals. Biologist Jonathan Birch (2021), whose testimony contributed to this amendment, provided a concise definition of sentience as an animal's capacity to have feelings with positive or negative quality. These included feelings of pain, pleasure, comfort, discomfort, boredom, excitement, contentment, frustration, anxiety and joy, making practices such as eyestalk ablation in shrimps an ethical and welfare issue.

Eyestalk ablation requires the removal of one or both of the eyestalks of a mature broodstock female shrimp in the belief that it will induce greater egg production (Shrimp Welfare Project, n.d.). This is done by crushing, cutting, burning or tying off, resulting in negative health, welfare and survival implications. Eyestalk ablation has been associated with several negative reactions indicative of pain, including a recoil reaction, flicking of their tails and rubbing the site of the wound (Birch et al., 2021). Studies on ablation-free approaches suggest that the avoidance of eyestalk ablation can actually lead to better reproductive performance from females and resilient offspring with lower mortality rates, and yet it continues in some regions (Zacarias et al., 2021).

While countries such as the United Kingdom recognise sentience in decapod crustaceans, most shrimp aquaculture occurs in regions that may not have such beliefs enacted in legislation. China, Thailand, Indonesia, India, Vietnam, Brazil, Ecuador and Bangladesh account for over half the global production of farmed shrimp to meet ever-increasing demand from the USA, Europe and Japan (World Wildlife Fund, 2025). As long as consumers in these countries demand cheaply produced shrimp, intensified aquaculture will persist to the detriment of these sentient animals.

Insects are another invertebrate animal increasingly subjected to intensive farming conditions. For decades, their small nervous systems and likely reflexive nocifensive behaviours created scepticism about the possibility of them feeling pain, but recent work has disputed these beliefs (Crump et al., 2023). When researchers Crump and colleagues (2023) reviewed over 300 studies to assess pain across a range of insects, they reported some startling results. Drawing on Birch and colleagues' (2021) eight neural and behavioural criteria for assessing the likelihood of pain, (namely, 1) nociception, 2) sensory integration, 3) integrated nociception, 4) analgesia, 5) motivational trade-offs, 6) flexible self-protection, 7) associative learning, and 8) analgesia preference), they concluded that several insects could likely feel pain. Strongest evidence was found for adult flies and cockroaches, followed by adult bees, wasps and ants. They commented that there were some evidence gaps given the lack of studies related to pain-relevant neurobiology and behaviour in most insects, but reminded the reader that: "Absence of evidence, though, is not evidence of absence" (p. 3).

If cockroaches can feel pain, this has implications for the many millions of them housed in factory farms in China. Ground up as part of a healing potion in traditional medicines or fed to livestock as a source of protein, the commodification of cockroaches fails to acknowledge any capacity to suffer or their importance in natural ecosystems (Millman, 2022). This lack of consideration may be compounded by the disgust factor for this resilient creature who can survive blasts of radiation, go for a month without food and remain submerged under water for an hour.

Farming edible insects requires considerably less space and resources compared to more traditional mammal and avian livestock, allowing millions of insects to be housed in one facility. Described as both leftovers of agricultural production and leftovers of animal ethics (Loo & Sellbach, 2013), these "mini-livestock" risk the same exploitation and welfare issues experienced by mammals, birds, reptiles, fish and decapod crustaceans. Small-scale insect rearing occurs in many Asian and African regions, where insects are locally sourced and disseminated through local markets, but industrial-scale edible insect production is growing rapidly in Western and Asian countries. Research among United Kingdom insect farmers identified that insects were often dismissed as outside moral obligation. This related to limited consensus about their sentience, and therefore, what, if any, ethical issues were associated with their confinement and slaughter, compounded by limited formal regulations prescribing standards for edible insect production (Bear, 2021). Insects may be killed by freeze-drying, sun-drying or boiling, all of which fail to meet criteria related to the avoidance of unnecessary distress or harm to a living being. The 2020 United Nations Food and Agriculture Organisation (FAO) publication *Guidance on Sustainable Cricket Farming* (Hanboonsong & Durst, 2020) describes harvesting of adult crickets as if they are inanimate produce rather than sentient animals:

> Once harvested, the crickets can be bagged alive immediately and sold to buyers. However, if the crickets are transported, heat generation in the bags could cause the release of an undesirable amount of histamine, which can cause allergic reactions in some people [...] Other post-harvest options include boiling the crickets for five minutes to sterilise them.
>
> *(pp. 15–16)*

Research on invertebrate capacities involving adult crickets reports inter-individual differences in the traits of aggressiveness, general locomotor activity and exploratory behaviour (Lambert et al., 2021). Isolation and deprivation of environmental stimulation and other social experiences during nymphal development accounted for some of these individual traits, suggesting crickets have an awareness beyond that of a commodity to be bagged alive for transport.

Bees also suffer in a mobile form of intensive animal agriculture, with their hives treated as commodities of pollination and shipped over greater distances than could occur naturally (Ellis et al., 2020). Each year around 80 per cent of the USA stock of commercial pollinator bees—approximately two million colonies—is transported to Californian almond orchards to meet global demand for almond milk and other almond products. Confinement during transportation can negatively affect the health and lifespan of individual bees as their colony needs are subjugated to human production requirements. Bee expert Lars Chittka (2022) described bees as smart creatures with emotional states evident in their unique behaviours and choice of flowers. His research provided evidence that bees must remember colours, scents, shapes and routes when foraging in the flower "supermarket", and turning them into transportable pollination commodities had long-term welfare and ethical implications.

CROSSROADS IN ANIMAL WELFARE

In 2002, an article in the *Nature* journal pointed a finger directly at humanity's escalating effects on the environment. The word "anthropocene" used in this article to link climate change and human accountability was unleashed. Derived from the Greek terms *anthropo*, meaning human, and *cene*, meaning new, it encompassed the way humans have become the dominant force in shaping the world, and therefore animals and ecosystems on the planet. While farmed animals increasingly bear the brunt of extreme weather events associated with climate change, they have also been labelled as a significant contributor to the problem through methane production and land clearance for grazing to feed ever-increasing human populations.

Ironically, exploitation of farmed animals has thus turned them into a potential risk to the world and human wellbeing.

Confining sentient animals within industrial farm systems is fundamentally wrong, and more and more it seems humanity—as well as other animals and the planet itself—is paying for this cruelty. Philosopher Melanie Challenger (2021) commented that: "[…] we don't know the right way to behave towards life. This uncertainty exists in part because we can't decide how other life forms matter or even if they do" (p. 2). As environmental uncertainties escalate around the world, it has become clear that other life forms do matter and must be taken into consideration if there is to be any hope for the future. The welfare of farmed animals and the removal of a system that disregards animal feelings and emotions as sentient beings must be addressed at both an individual and societal level. As long as there is demand for meat and animal products, animals will continue to be exploited for profit. Reducing that demand by changing attitudes and behaviours will increase incentives to find alternatives, but there is still a long way to go to achieve this logical outcome.

Nonhuman and human animals are at crossroads, and it is the humans who have the power to determine which path to take. Look into the eyes of a farmed animal now living in a farmed animal sanctuary and the choice should be easy—these sentient animals are more than production items within an industrialised farm system. They are unique individuals, and to perceive them otherwise is fundamentally wrong. All the animals who shared their stories in the preceding chapters formed meaningful connections with other animals and humans once free of the traumas from their past lives. Those left behind in abhorrent circumstances continue to suffer in silence and are reliant on the small number who escape to create empathy and compassion for their suffering.

A growing number of humans are starting to acknowledge the wrongs inflicted on farmed animals, and they are stirring up a groundswell of change. Those reptiles, fish, shrimps and insects lacking the cute factor of mammals and birds are also in need of compassion and change as they remain increasingly vulnerable to human production demands. The current industrialised animal agriculture system is broken, and increasing production efficiencies or eliminating an animal's capacity to feel negative affective states is not going to solve a growing global animal welfare crisis.

REFERENCES

Animal Charity Evaluators (2020). *Farmed fish welfare report – 2020*. https://animalcharityevaluators.org/research/reports/farmed-fish-welfare-report/#indicators-farmed-fish-welfare

Balcombe, J. (2016). *What fish knows: The inner lives of our underwater cousins*. Scientific America/Farrar, Strauss and Giroux.

Bear, C. (2021). Making insects tick: Responsibility, attentiveness and care in edible insect farming. *EPE: Nature and Space*, 4(3), 1010–1030. https://doi.org/10.1177/2514848620945321

Beilharz, R. G., & Zeeb, K. (1981). Applied ethology and animal welfare. *Applied Animal Ethology, 7*(1), 3–10. https://doi.org/10.1016/0304-3762(81)90047-X

Bekoff, M. (2024). *The emotional lives of animals: A leading scientist explores animal joy, sorrow, and empathy——and why they matter.* New World Library.

Birch, J. (2021). *Invertebrate sentience*. [Video]. tps://bit.ly/birchWBI

Birch, J., Burn, C., Schnell, A., Browning, H., & Crump, A. (2021). *Review of the evidence of sentience in Cephalopod Molluscs and Decapod Crustaceans*. London School of Economics and Social Science. https://www.lse.ac.uk/news/news-assets/pdfs/2021/sentience-in-cephalopod-molluscs-and-decapod-crustaceans-final-report-november-2021.pdf

Challenger, M. (2021). *How to be animal. A new history of what it means to be human*. Canongate Books Ltd.

Chittka, L. (2022). *The mind of a bee*. Princeton University Press.

Crump, A., Gibbons, M., Barrett, M., Birch, J., & Chittka, L. (2023). Is it time for insect researchers to consider their subjects' welfare? *PLoS Biology, 21*(6), e3002138. https://doi.org/10.1371/journal.pbio.3002138

Darlim, S. (2024, November 18). Farmed animals: How factory farming affects different species. https://animalcharityevaluators.org/blog/farmed-animals/

Ellis, R.A., Weis, T., Suryanarayanan, S. & Beilin, K. (2020). From a free gift of nature to a precarious commodity: Bees, pollination services, and industrial agriculture. *Journal of Agrarian Change, 20*(3), 437–459. https://doi.org/10.1111/joac.12360

Fischer, B. (2020). In defense of neural disenhancement to promote animal welfare. In L. Johnson, A. Fenton, & A. Shriver (Eds.), *Neuroethics and non-human animals. Advances in neuroethics* (pp. 135–150). Springer. https://doi.org/10.1007/978-3-030-31011-0_8

Guarcello, S., Markowitz, L.C. & Williams, J. (July/August 2023). Who's the real predator? A look into alligator farming. *Florida Bar Animal Law Section, 97*(4). https://flabaranimals.org/whos-the-real-predator-a-look-into-alligator-farming/

Hanboonsong, A. & Durst, P. (2020). *Guidance on sustainable cricket farming – A practical manual*. Bangkok, FAO. https://doi.org/10.4060/cb2446en; https://www.fao.org/3/cb2446en/cb2446en.pdf

Lambert, H., Elwin, A., & D'Cruze, N. (2021). Wouldn't hurt a fly? A review of insect cognition and sentience in relation to their use as food and feed. *Applied Animal Behaviour Science, 243* (October). https://doi.org/10.1016/j.applanim.2021.105432

Learmonth, M. J. (2020). The matter of non-avian reptile sentience, and why it "matters" to them: A conceptual, ethical and scientific review. *Animals, 10*(5), 901. https://doi.org/10.3390/ani10050901

Leshko, I. (2019). *Allowed to grow old. Portraits of elderly animals from farm sanctuaries*. University of Chicago Press.

Loo, S., & Sellbach, U. (2013). Eating (with) insects: Insect gastronomies and upside-down ethics. *Parallax, 19*(1), 12–28. https://doi.org/10.1080/13534645.2013.743290

Millman, O. (2022). *The insect crisis: The fall of the tiny empires that run the world*. W.W. Norton & Company, Inc.

PETA (2025). *Exposed: Crocodiles and alligators factory-farmed for Hermès 'luxury' goods*. https://investigations.peta.org/crocodile-alligator-slaughter-hermes/

RSPCA (2025). What is the RSPCA's view on crocodile farming? *RSPCA Knowledgebase*. https://kb.rspca.org.au/knowledge-base/what-is-the-rspcas-view-on-crocodile-farming/

Shrimp Welfare Project (n.d.). *Eyestalk ablation-free*. https://www.shrimpwelfareproject.org/eyestalk-ablation-free

Singh, G.G., Sajid, Z. & Mather, C. (2024). Quantitative analysis of mass mortality events in salmon aquaculture shows increasing scale of fish loss events around the world. *Scientific Reports, 14*, 3763. https://doi.org/10.1038/s41598-024-54033-9

Vindas, M.A., Johansen, I.B., Folkedal, O., Höglund, E., Gorissen, M., Flik, G., Kristiansen, T.S., & Øverli, Ø. (2016). Brain serotonergic activation in growth-stunted farmed salmon: adaption versus pathology. *Royal Society Open Science, 3*(5), 160030. https://doi.org/10.1098/rsos.160030

World Wildlife Fund (2025). *Farmed shrimp*. https://www.worldwildlife.org/industries/farmed-shrimp

Zacarias, S., Fegan, D., Wangsoontorn, S., Yamuen, N., Limakom, T., Carboni, S., Davie, A., Metselaar, M., Little, D.C., & Shinn, A.P. (2021). Increased robustness of post larvae and juveniles from non-ablated Pacific white leg shrimp, Penaeus vannamei, broodstock post challenged with pathogenic isolates of Vibrio parahaemolyticus (VpAHPND) and white spot disease (WSD). *Aquaculture, 532*, 736033. https://doi.org/10.1016/j.aquaculture.2020.736033

Index